Synthetic Data for Official Statistics

A Starter Guide

UNITED NATIONS

Geneva, 2022

ECE/CES/STAT/2022/6

Sales no.: E.22.II.E.29
ISBN: 978-92-1-117312-3
eISBN: 978-92-1-002170-8
ISSN: 0069-8458

Preface

Synthetic data provides new opportunities for National Statistical Offices (NSOs) to maximise the amount of statistical information that data users can utilise, while keeping statistical disclosure risks at a minimal level.

NSOs have always faced a tension in deciding how much information to release to users, and how much emphasis to place on minimising the risk of disclosure of the details from a specific record within an underlying microdata set (which could for example relate to a specific individual). This tension has increased in recent years, as NSOs have faced greater pressure to release more detailed data, and faster than ever before.

Traditionally, NSOs have managed these risks either via the public dissemination of only *tabulated aggregates* of the underlying micro-datasets, and/or by authorising certain groups of users (e.g., accredited researchers) access to some of their microdata (i.e., accessing individual records to perform sophisticated analyses). However, these are not ideal solutions, because:

- Tabulated aggregates will often not satisfy the information needs of many users, who may demand additional breakdowns of tabulated data. Each new breakdown provided makes it harder to suppress the risk of disclosing information about a specific individual record, and requires more resources to monitor and manage the production of such tables.

- In the case of statistical microdata access, especially stringent measures are required to manage which users can access such datasets. The process and procedures to manage access can be cumbersome, bureaucratic, time-consuming and not without risks that could potentially have serious consequences in the event of a disclosure.

Into this arena comes the application of **synthetic data** as an alternative option for managing the release of data by National Statistical Offices, which may be more convenient for certain use cases, for which synthesised micro-level records may be sufficiently realistic to satisfy the analytical requirements of the users of such data, while posing a substantially reduced risk of disclosing information about the *original* data from which the synthetic data was synthesised.

Of course, the term "sufficiently realistic" is highly dependent on the particular use case for which such synthetic data is to be utilised by the user, and the extent to which the disclosure risk is acceptable to the NSO is also dependent on the specific circumstance in question. There are also a variety of different methods with which synthetic data sets can be created, each having its own advantages and disadvantages.

This guide is for those working in NSOs who are involved in managing access to statistical data, and who wish to explore the possibility of using synthetic data as a possible method for users to access it. The guide highlights some recent successful applications of synthetic data by a number of different NSOs, and introduces some of the different approaches that can be taken to creating synthetic data, including recommendations on which approaches to use in different situations, as well as practical tips and resources for getting started for practitioners.

There are also chapters dedicated to disclosure risk considerations when releasing synthetic data (including privacy preserving techniques, and measures to assess disclosure risk), and on utility measures that can be used to assess how well the synthetic data meets the analytical needs of users.

This guide is based on the results from the UNECE High-Level Group on Modernisation of Official Statistics (HLG-MOS) Synthetic Data Project (2020-2021) and earlier work of the Blue Skies Thinking Network activity of synthetic data, and is approved by the HLG-MOS.

We hope that this guide helps you on your journey towards implementing synthetic data in your organization!

Acknowledgments

The Introduction was authored by **Kate Burnett-Isaacs** and **Claude Girard** at Statistics Canada.

The chapter on uses of synthetic data was authored by **Kate Burnett-Isaacs**, Statistics Canada.

The chapter on methods for creating synthetic data was authored by **Kenza Sallier**, Statistics Canada and **Gillian Raab**, Emeritus Professor, Edinburgh Napier University and Part-time Research Fellow at the Administrative Data Research Centre Scotland.

The chapter on disclosure considerations for synthetic data was authored by: **Kate Burnett-Isaacs** and **Claude Girard** at Statistics Canada; **Gillian Raab**, Emeritus Professor, Edinburgh Napier University and Part-time Research Fellow at the Administrative Data Research Centre Scotland; **Alistair Ramsden**, Statistics New Zealand; **Manel Slokom**, Delft Technical University, Radboud University and Statistics Netherlands; **Christine Task**, Knexus Research and **Mustafa Salamh**, **Alison Wardlaw**, **Ehssan Ghashim**, **Rachel Ostic** and **Phil Belanger** at the Canadian Revenue Agency.

The chapter on utility measures for evaluating synthetic data was authored by: **Gillian Raab**, Emeritus Professor, Edinburgh Napier University and Part-time Research Fellow at the Administrative Data Research Centre Scotland; **Manel Slokom**, Delft Technical University, Radboud University and Statistics Netherlands, and staff at the Innovation and Analytics Division of the Canadian Revenue Agency.

Contributions to this guide were also made by **Gail Bouchette**, **Jia Xin Chang**, **Héloïse Gauvin**, **Claude Girard** and **Steven Thomas** at Statistics Canada; **Ioannis Kaloskampis**, United Kingdom Office for National Statistics; **Alistair Ramsden**, Statistics New Zealand; **Rolando Rodriguez**, U.S. Census Bureau; **Manel Slokom**, Delft Technical University, Radboud University and Statistics Netherlands; **Christine Task**, Knexus Research; and **Mustafa Salamh**, **Alison Wardlaw**, **Ehssan Ghashim**, **Rachel Ostic** and **Phil Belanger** at the Canadian Revenue Agency.

Content presented during the HLG-MOS Synthetic Data Project during 2020 and 2021 has been included in this guide from: **Geoffrey Brent** and **Joseph Chien** at the Australian Bureau of Statistics; **Joerg Drechsler**, Institute for Employment Research; **Nicolas Grislain**, Sarus; **Ioannis Kaloskampis**, Office for National Statistics; **Øyvind Langsrud**, Statistics Norway; **Gillian Raab**, Emeritus Professor, Edinburgh Napier University and Part-time Research Fellow at the Administrative Data Research Centre Scotland; **Rolando Rodriguez**, U.S. Census Bureau; **Kenza Sallier**, Statistics Canada; and **Christine Task**, Knexus Research.

This guide has been edited and prepared for publication by **Christopher Jones**, **Taeke Gjaltema** and **Wai Kit Si Tou** at the United Nations Economic Commission for Europe.

Contents

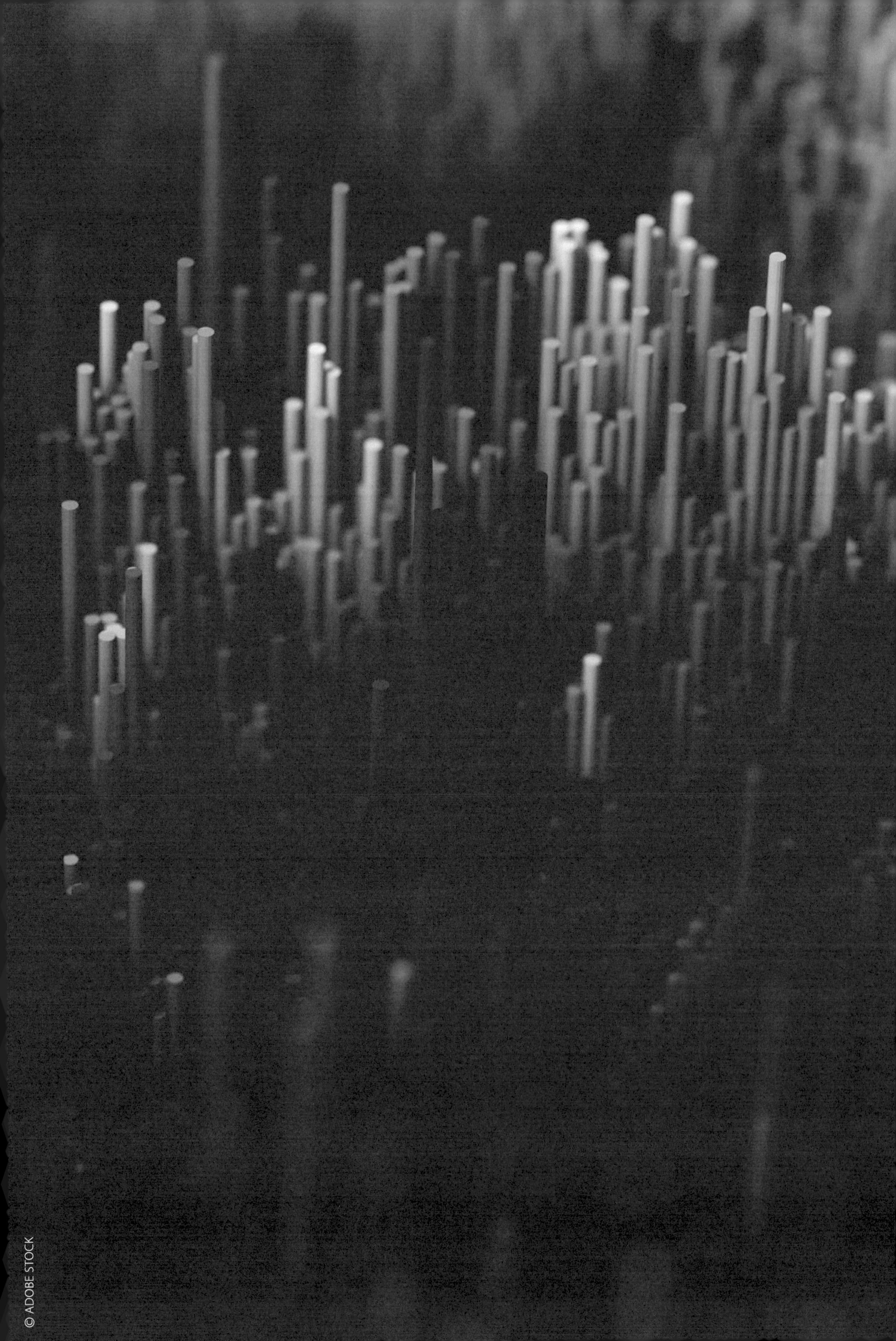

Chapter 1: Introduction

Data are a valuable resource, providing critical input for statisticians, economists, and data scientists, to generate timely and granular insights that respond to the information needs of a broad range of stakeholders. In a world where increasingly large volumes of data are coming from an increasing number of providers, National Statistical Offices (NSOs) are using innovative approaches to maintain data standards and definitions, good privacy and confidentiality management systems, and responsible data-sharing.

NSOs have a leadership role to play in establishing safe and transparent ways to share data, expertise, and best practices to support the use of data for testing, evaluation, education, and development purposes. With data integrity and confidentiality at the forefront, NSOs are well-positioned to provide the tools, methods and approaches to promote responsible data-sharing in order to meet a growing number of stakeholders' needs in this ever-changing and fast-paced data ecosystem.

NSOs recognise that the call for greater openness and transparency of data must be met while simultaneously remaining steadfastly committed to protecting the confidentiality and privacy embedded in their data holdings. The dual mission of NSOs is nicely conveyed by Duncan *et al.* (2011) who write on page 12:

> *Data stewardship organization is serving two masters* [providing high-quality information and protecting confidentiality] – *each with conflicting interests and concerns.*

It is widely acknowledged that releasing information that is both useful and completely safe cannot be achieved in full.[1] The Office of the Privacy Commissioner of Canada writes that:[2]

> *It should be noted that there is no such thing as zero* [disclosure] *risk when releasing data.*

Also, it is widely acknowledged that 'safety' is relative, not absolute. For instance, Desai *et al.* (2016) write that:

> *'Safety' is a measure, not a state. For example, 'safe data'* […] *does not mean that the data is non-disclosive. 'Safe data' could be classified using a statistical model of re-identification risk, or a much more subjective scale, from 'very low' to 'very high'. The point is that the user has some idea of 'more safe data' and 'less safe data'.*

It is within this context that NSOs must establish measures for the relative importance of the usefulness[3] of the statistical information released and the protection for the person-level (or business level) information gathered from which it is derived.

---

1   For instance, page 135 of El Emam (2013), or item 4 on page 5 of Elliott et al. (2016).

2   See paragraph 130 of https://www.priv.gc.ca/en/opc-actions-and-decisions/investigations/investigations-into-federal-institutions/2018-19/pa_20191209_sc/

3   In this context, usefulness relates to how well the (synthetic) data meet the analytical needs of users, as defined by the concept of utility, described in the next section.

With the emergence of synthetic data, NSOs now have a promising data release option that responds to the call to expand the usefulness of data holdings, while providing safeguards for the confidentiality of record-level information. Use of synthetic data presents the opportunity to move toward standard approaches with this starter guide to synthetic data in NSOs. This synthetic data guide comprises a compendium of theoretical methods to create synthetic data, a consensus regarding practical applications and best practices to promote consistency, transparency and comparability within statistical agencies, as well as for users in academia and the private sector. This guide is intended to provide practical and direct guidance to decision-makers working in NSOs to help them to determine if synthetic data is the right solution for them to facilitate responsible data-sharing.

This guide opens with the most common scenarios for NSOs where synthetic data would be a suitable solution (Chapter 2). Chapter 3 discusses methods used to generate synthetic data, as well as recommendations for determining suitable applications for them. Chapter 4 highlights important considerations when releasing synthetic data, including privacy preserving techniques, and measures to assess disclosure risk. Finally, Chapter 5 presents utility measures that can be used to assess how well the synthetic data meets the analytical needs of users.

1.1   Key concepts for synthetic data

Any discussion about synthetic data must involve an understanding of utility as well as privacy, sensitivity, security and confidentiality, since decisions made about the creation of synthetic data will involve balancing some of these concepts. To explore these ideas further, it helps to refer to Figure 1, which illustrates a linear representation of the life cycle of data holdings, inspired by the 4G framework[4] (Rancourt, 2019): Data are gathered, guarded, grown and given.

Figure 1   Illustration of how data is collected, transformed, creates value
and then disseminated based on the 4G framework

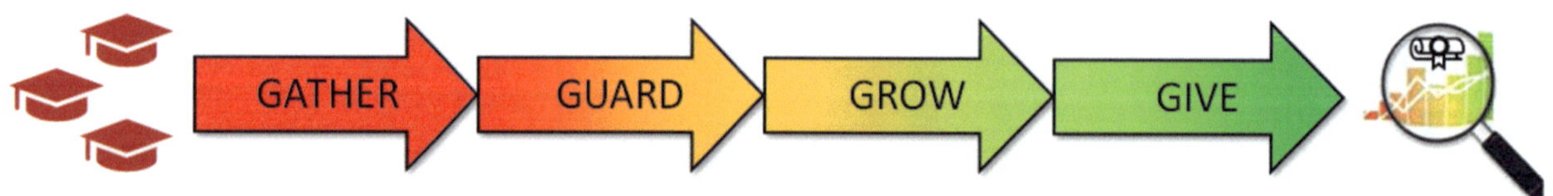

Gather:

Privacy is associated with how personal data are gathered by NSOs, involving such considerations as the right of individuals to be free from observation, the NSO's entitlement to ask for and obtain information pertaining to individuals, and individuals' consent to share their information with a NSO only under agreed-upon terms.[5]

---

4   Some representations include a fifth 'G' with Governance overseeing the other four Gs described here.

5   No consensus exists on the precise meaning of 'privacy': not only may definitions differ from one field to another (e.g., computer science vs official statistics), but they may also vary within a given field (e.g., between NSOs).

Guard:

Once information has been entrusted to an NSO through a sharing agreement, it exists in a highly identifiable person-level form, and must therefore be guarded against any unauthorised access. This is where the various data-storing options, access protocols and security measures available to NSOs come into play.

Grow:

Utility is then grown by transforming person-level data into statistical information, a form better suited for release purposes. This is where finite population estimates are calculated, complex statistical analyses are performed, and analytical data sets are produced. For example, data gathered from graduates could be transformed into enrolment numbers by major fields of study.

Give:

Finally, the data are given to users. Here, confidentiality issues relate to unwarranted disclosure, as per the information-sharing agreement, of the personal data entrusted to a NSO that may occur when statistical information is released.

Thus, the 4Gs described above imply that privacy and confidentiality are distinct notions, as they arise at opposite ends of the data life cycle (Gather and Give). This is not to say these notions are unconnected, but rather that concerns one may have with regard to each deserve separate consideration.

The concepts of disclosure and disclosure risk extend the discussion on confidentiality. Disclosure risk is the risk or possibility of inappropriate release of data or attribute information (OECD, 2003), and applies to the dissemination of any aggregate statistic or microdata set, including synthetic data. In fact, there is often a balance when creating synthetic data between the utility and the disclosure risk, as the more closely synthetic data results emulate those of the original data, the higher the risk that confidential information in the original data could be disclosed.

In the event of unwarranted disclosure, it is reasonable to expect that the harm done to an individual will be greater the more sensitive the information revealed is. However, this does not mean that information that is less sensitive is any less confidential, since that information is confidential due to being covered by the terms and conditions of the sharing agreement, and not because of how sensitive it is deemed to be by its custodian.

To help manage confidential data, NSOs can rely on the Five Safes framework,[6] developed at the Office for National Statistics in the United Kingdom, notably when thinking of data access solutions (e.g., Desai *et al.* (2016)). One of the framework's five dimensions is Safe Data, which examines the disclosure risk posed by the data itself. Both synthetic data sets and Public-Use (Microdata) Files (PUFs or PUMFs) are access solutions that are made to score high on the Safe Data scale through the use of appropriate disclosure control methods.

____

6   See www.fivesafes.org for details.

1.2   Data access options of National Statistical Offices

Typically, much of the data collected and stored by NSOs is sensitive in nature, and can only be accessed by trusted employees and researchers working in a secure environment under strict conditions. Anonymised versions of original data files are not good candidates for public release because the information they contain can often still be attributed to a respondent, constituting a violation of the requirements of statistical legislation, regulations, policies, standards, and relevant ethical guidelines.

NSOs have a number of alternatives for safe data sharing, ranging from dummy files to Research Data Centres; however, there remains a gap in sustainable access solutions that balance utility with confidentiality.

The core business of NSOs is to disseminate aggregate statistical information. This information is very valuable to the public, as well as industrial and political decision makers. However, due to confidentiality reasons, aggregate information does not provide the level of detail in the data that users of official statistics are increasingly looking for.

NSOs may also have Public Use Microdata Files (PUMF) at their disposal, to provide users with more granular data. PUMFs are anonymised microdata files containing information relating to a sample of individual units from a survey, census or administrative file. The anonymisation process is one where identifying information is modified or suppressed to avoid identification of individual entities in a data file. Though PUMFs provide more granularity, the anonymisation process can limit their analytical value.

To provide the confidential microdata to users, NSOs have created physical Research Data Centres (RDCs) to facilitate research projects that draw, for example, on survey or administrative microdata. RDCs provide direct access to a wide range of anonymised but not fully confidentialised microdata to accredited researchers under strict conditions.

These physical location requirements can pose constraints for users. As a result, some NSOs such as Statistics Canada, Statistics New Zealand, the Australian Bureau of Statistics and Statistics Netherlands, have introduced real-time remote access or remote execution solutions, to enable users to quickly obtain a full range of descriptive statistics without the physical location requirement. This access option is typically limited to accredited researchers.

Finally, and least usefully, NSOs may share dummy files, which are data sets where almost none of the original data set's analytical value is preserved, and the focus is on maintaining the structure and the logical rules of the original file. These files provide very little analytical value and so are not useful to many users.

These traditional options, used by NSOs for data release, are depicted in Figure 2, with dummy files on the far left being the dissemination option that carries the least disclosure risk (but also the least utility), to the original data on the top right, where the utility is the greatest (but with the highest disclosure risk).

Figure 2   Confidentiality versus utility of current data release mechanisms

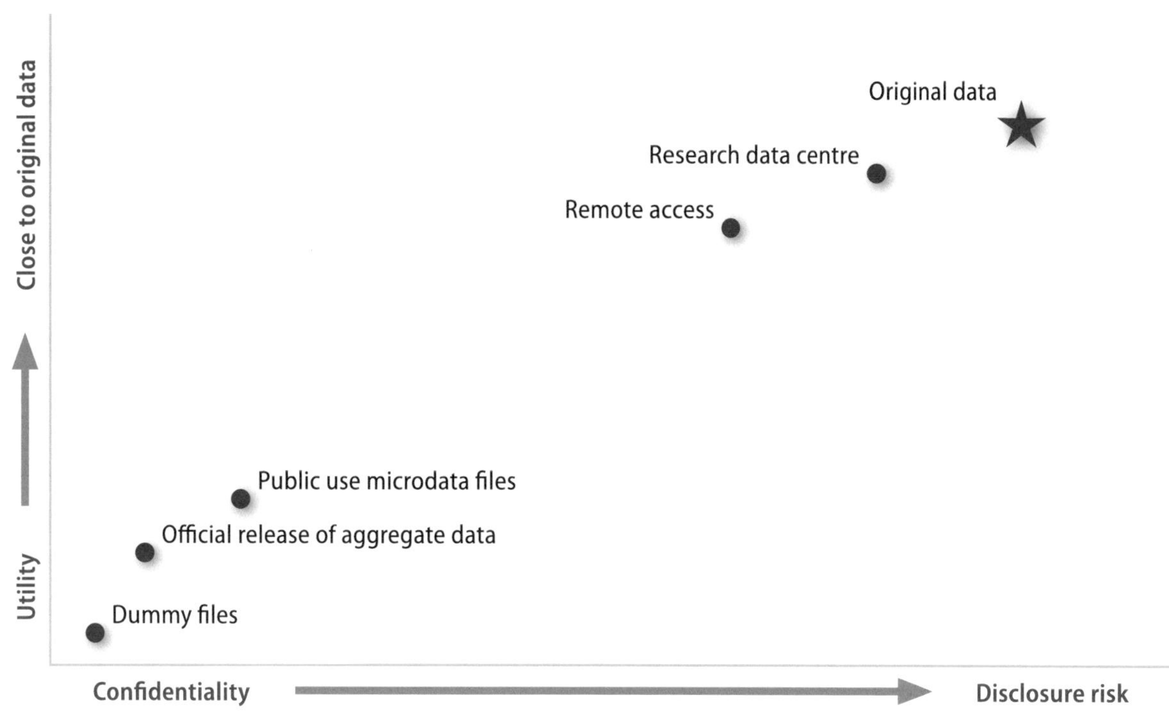

As shown in Figure 2, particularly between PUMF and remote access disclosure options, NSOs face a gap in utility and confidentiality options that needs to be filled in order to be more transparent, and to ensure data holdings are made more accessible to users. Synthetic data is a viable alternative data dissemination strategy that can facilitate data access, especially in cases involving highly sensitive data.

1.3　A brief introduction to synthetic data

Synthetic data is defined as being stochastically generated data that has analytical value, and which maintains high levels of disclosure control. Synthetic data has its roots in data editing and imputation methods, and has become more developed with recent advances in computing and data science methods, as well as a drive by NSOs for more open and transparent data sharing. Creating synthetic data involves a generation or modelling process that targets both the preservation of analytical value and confidentiality.

The advantage that synthetic data brings to the suite of disclosure options for NSOs is that it breaks the direct link between collected data and the released outputs, due to the modelling or generation process.

Figure 3 illustrates the synthetic data generation process. At a high level, the goal of synthetic data is to take the original data set (D) that outputs results (Θ(D)), and to synthesise the data (D') so that the confidentiality of the records is maintained, while also ensuring that the results of the synthetic data (Θ(D')) match as closely as possible to those results obtained based on the original data.

Figure 3　Illustration of the synthetic data generation process

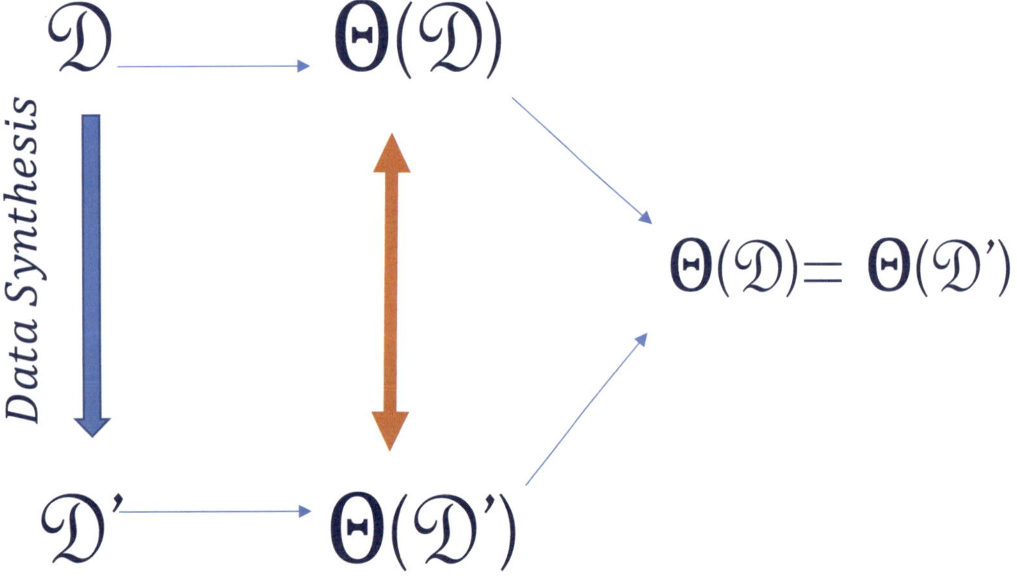

Source: Sallier (2020).

1.4   Types of synthetic files

There are various types of synthetic files, and elements of each should be taken into consideration when determining which one is most suitable for its purpose.

1.4.1   Dummy files

Dummy files are data sets where almost none of the analytical value is preserved. The focus is on maintaining the structure and the logical rules of the original file. These files are often used to test programs and processes, and to provide remote access to structurally similar (but not inferentially valid) data sets. Since there is no analytical value to these files, there is also almost no disclosure risk. This type of file is well known and in wide use, so is not a focus of this guide.

1.4.2   Fully synthetic files

In fully synthetic data files, all of the variables are synthesised. The goal is to preserve significant levels of relevant analytical value compared to the original data set, in order to meet the needs of the user. For example, this could be done by preserving within the synthetic data the univariate distributions for the variables from the original data, or it could be done by preserving one or more multivariate or joint distributions of those variables. Variables could be generated in order to preserve only particular statistics (e.g., margins, mean, etc.), or entire sets of relevant descriptive statistics for relevant distributions.

These files have the same use as PUMFs, but present greater analytical value when the joint distribution of the original data is preserved. These files strive to present low disclosure risk, however they could present inferential disclosure risk (or merely perceived disclosure risk). When their utility increases, so too can their potential disclosure risk, and so these objectives must be weighed against each other when creating such data sets.

1.4.3   Partially synthetic files

In partially synthetic files, only some of the variables are synthesised. The goal is the same as the approach with fully synthetic files, but the approach typically focuses on a subset of variables in the data set. For example, one could synthesise the most sensitive variables and leave all of the other variables untouched.

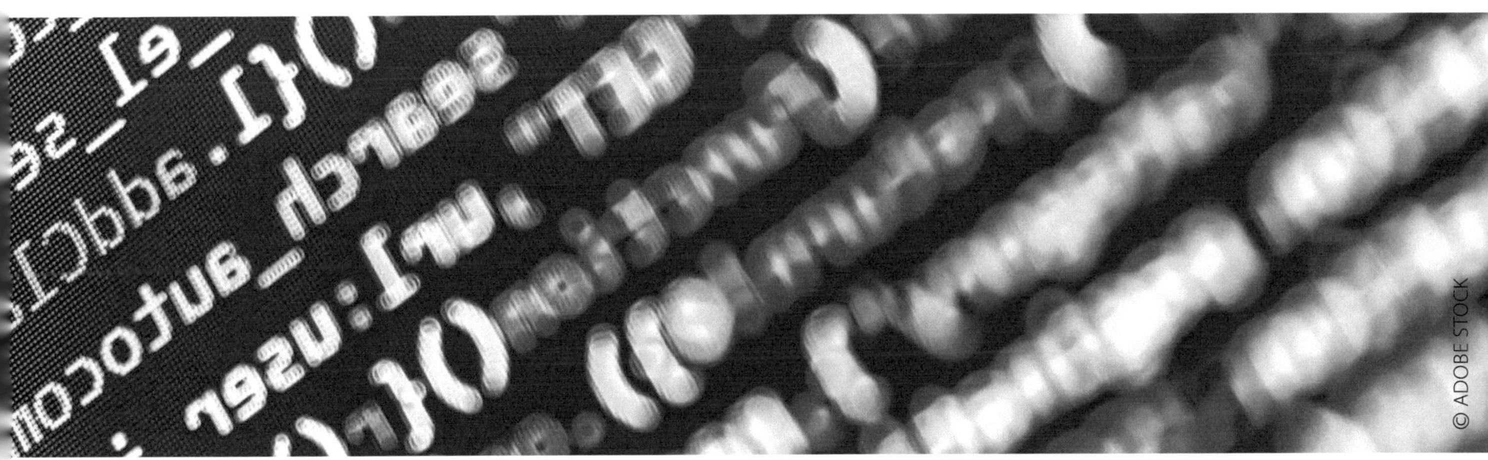

Chapter 2:   Uses of synthetic data

Synthetic data can solve statistical disclosure problems faced by NSOs, but the value of synthetic data varies with the nature of the problem faced. This section explains the main uses of synthetic data in NSOs, with a discussion of utility, disclosure risk requirements, and risk mitigation.

2.1   Releasing synthetic microdata to the public

Traditional data release approaches used by NSOs can limit users' access to high-quality microdata. However, with the increasing emphasis on transparency and data access, NSOs are exploring synthetic data as a new data release option.

If synthetic microdata are publicly disseminated, NSOs cannot know or control how the data are used. Therefore, there is no prior knowledge of what distributions, variables, or relationships need to be preserved in the synthetic data, and so in principle as many as possible of the original data's relationships should be preserved to maximise the utility of the released file. However, this increased utility also tends to increase disclosure risk, and since the audience is the public, there are no other controls or vetting processes over the access to, and use of, the data. Therefore, the confidentiality requirements in such a scenario are of utmost importance.

2.1.1   **Example:** Statistics New Zealand's synthetic unit record files

Statistics New Zealand is now releasing more granular data, and one way they are doing this is through Synthetic Unit Record Files (SURFs). SURFs are generated by a mathematical model, based on, but not the same as, the original data. Statistics New Zealand has released a few such files, including one based on their 2007 income survey and a 'Census for Schools' SURF based on their 2019 household savings survey and census.

These files are semi-realistic representations of a sample of the New Zealand population, but respecting only the distributions, variables and relationships that are preserved by the synthesis method that was used. Published data are typically perturbed (noise-added) representations of population data. Hence, privacy of the data is preserved because even if some of the synthetic data resembles the original data, an attacker (someone who deliberately seeks to disclose or breach confidentiality rules) arguably cannot be sure which data are the same, which data are different, and whose data is whose.

For example, the SURF files released in 2007 were based on the income survey from the second quarter of 2003. Each file contains over 11,000 records, and one hundred such files were released in 2007, representing 100 samples (some potentially overlapping) of the New Zealand population between the ages of 25 and 64, who participated in paid work. The variables included are age, sex, ethnicity, highest educational qualification, weekly hours worked and weekly income.

These SURFs were released with a clear recommendation that they could be used for the purposes of teaching or learning, developing analytical methods or processes, or some level of statistical inference (New Zealand Income Survey Super SURF, n.d.). The policy of Statistics New Zealand is to release SURF data with appropriate metadata about their methodology, inferential validity, and safety (disclosure risk).

2.2   Testing analysis

Some NSOs grant confidential microdata access to trusted parties, remotely or at physical Research Data Centres (RDCs), but going through security checks, vetting and approvals can greatly delay important research and analysis projects. Synthetic data could be useful in this context by allowing researchers to more easily develop and test their models, algorithms or analyses, and potentially to conduct exploratory data analysis and/or to determine initial hypotheses or conclusions, while they wait for access to the original data. The original data would be required only to complete their research, requiring fewer journeys to be made to the RDC premises.

In some ways, such a use case is easily accommodated, because NSOs typically know the types of analyses that researchers would conduct. NSOs can generate synthetic data that preserve specific distributions, variables and relationships of interest to the researchers, while variables or relationships that are not of interest need not be preserved, allowing for more flexibility in synthetic data modelling choices. However, such use cases may involve extensive work for NSOs to provide bespoke synthetic data files for users.

2.2.1   **Example:** Statistics Canada synthetic census-based data

Starting in June 2021, one of the ongoing projects of Statistics Canada has been the creation of a synthetic version of a census-based database. Its objective is to test and run the new dynamic micro-simulation model of the Canadian retirement and income system, built for Employment and Social Development Canada (ESDC). A desired feature of the model is non-confidentiality, allowing the model to be used at any location. This would increase operational flexibility and the potential for external collaboration and broader use. Different options were explored, and synthetic data was chosen as it seems likely to provide results closest to the original data.

The original database represents a part of the Canadian population in 2011, with some basic cross-sectional characteristics of the starting population. Based on this, the dynamic micro-simulation can be thought of as experimenting with a virtual society of millions of individuals whose lives evolve over time. This micro-simulation model will allow academics, researchers and government policy makers to model changes to the Canada Pension Plan (CPP), enabling research on public pensions and, more broadly, income security in retirement. The public (synthetic) database, of roughly 8.6 million records, will support model development, as well as preliminary programme assessment, policy analysis and research. For final analysis and publication, the micro-simulation model will be run on the original data housed in the Research Data Centres.

2.2.2   **Example:** Provision of synthetic data for users of the Scottish Longitudinal Study

The Scottish Longitudinal Study (SLS) is a source of linked data. At the core of the SLS is a 5% sample from the census data for Scotland, where individuals (SLS members) are linked over time between censuses. The data also include information on all household members of each SLS member. Further data sets are permanently linked to the SLS, including births, deaths, marriages and school-record data. Other data sets can be linked to the SLS for specific projects related to hospital admissions, cancer registrations and many other topics.[7]

Currently, an extract of the data is prepared for each user with an approved project. The extract can only be viewed and analysed under supervision in the National Records of Scotland Offices in Edinburgh, which incurs a travel burden on researchers. To reduce this burden, researchers can request a synthetic data extract at the time of applying for access to the original data. To receive synthetic data, researchers and other members of the research team must complete safe-researcher training and agree to comply with conditions on the storage and use of synthetic data. The synthetic extract is then supplied to the researcher to analyse on their own computer. Results from synthetic data can only be shared among members of the research team and no findings from the synthetic data can be published. Final analyses for publication must be run on the original data in the Edinburgh office. In exceptional cases, SLS staff can run analyses remotely.

2.2.3 Example: Using synthetic data to test machine learning algorithms at the Australian Bureau of Statistics

Machine learning (ML) and artificial intelligence techniques have become more prevalent both for producing and analysing official statistics. At the Australian Bureau of Statistics (ABS), a need arose for data scientists to test ML methods using published (and non-confidential data) that represented entities and relationships of interest for their ML models, such as persons, households, regions, industries and business units.

Using a micro-simulation model, the ABS was able to create a synthetic data set, using only publicly available information that provided the details and relationships appropriate for testing these models.

2.3   Education

High-quality data is needed in order for students, academics, and users in general, to learn new concepts and methods related to a variety of topics, such as data science, statistics, data analysis and even technology. The more complex the methods (such as machine learning or complex statistics), the more important it is that the data yield realistic results.

In providing data for such educational purposes, NSOs may know the specific method or topic that is being studied and, as in the testing analysis use case, may preserve only the distributions of interest. Alternatively, synthetic data made for another use could be repurposed for educational use, provided that the use case in question has similar utility and disclosure risk requirements to the original purpose for which that synthetic data set was created.

While for many other use cases, the requirement for high utility limits options for minimising disclosure risk, for educational and training use, the confidentiality requirements can vary greatly: in some NSOs, as soon as microdata leaves the premises, the data must meet top disclosure standards no matter who is using the data. In other cases, the students may have some level of security clearance or agreement with the NSO already, therefore lowering the confidentiality requirements for the synthetic data.

2.3.1   Use Case: The Canadian Health Measures Survey

Many universities in Canada are developing undergraduate programmes that aim to develop, among other things, capacities in working with large data sets.

The Canadian Health Measures Survey (CHMS) includes a comprehensive data set from a questionnaire, as well as physical and laboratory data. The data are currently accessible only via Research Data Centres (RDCs), which require stringent security checks that can take time to complete. For undergraduate research programmes to work with CHMS data, it is proposed to create scientific use files (SUF) for which some of the variables and survey weights are synthesised to allow open access to the data. This data set would allow students to develop analytical skills on data derived from a complex survey design, however the publication of results derived from SUFs would be prohibited, as users would require access the original microdata for publication purposes.

2.4   Testing of technology

For the purposes of testing new software and technology, dummy data that represent the file layout and error rates of original data are often used. However, dummy data files have no analytical value, and as complex technologies such as artificial intelligence and machine learning become more prevalent, such testing increasingly requires more analytically realistic data. In these cases, synthetic data with some inferential validity can be beneficial.

The utility of the original data needs to be preserved to some extent, so that the results of the system can be assessed and verified, even though the conclusions drawn from the results have minimal value.

2.4.1 **Example:** Census systems testing at the Office for National Statistics (ONS)

The ONS used synthetic data in testing its census processing system prior to the 2021 census. Their census processing comprised several phases, each dependent on the previous, with various quality 'gate checks'. While many system tests worked on 'dummy' data, other tests required variable distributions that were representative of the population, to realistically verify that the checks worked as expected, and to ensure system resources were sufficient under realistic workloads. Dummy data were not sufficient in this use case, as specifications on required test data distributions ran to roughly 60 pages.

Multiple synthetic data sets were generated to test load balancing and the various functions used in the processing pipeline. Some synthetic variables that were not previously captured in the census were modelled and generated.

2.5   Tips to get started

When deciding whether synthetic data is the right solution for a given data release scenario, a firm understanding is needed of both the constraints involved to ensure confidentiality, as well as the requirements that the users will have for the synthetic data. While synthetic data can provide confidential data with high levels of analytical value for users, the level of confidentiality and the specific sort of analytical value that needs to be retained, is dependent on the use case in question. Table 1 summarises the key considerations for the different use cases presented in this chapter, along with the average balance of utility and confidentiality for each. The specific use case, together with these considerations play a role in determining which method to use to generate synthetic data.

Table 1   Summary of use cases, their key considerations and their balance of confidentiality and utility

Use Case	Key Considerations	Confidential/Utility Balance
Releasing microdata to the public	The synthesiser does not know who or how the data will be used.	High confidentiality as well as high utility are required.
Testing analysis	Specific analysis or variables distributions that must be maintained *may* be known at time of synthesis.	High confidentiality as well as high utility are required.
Education	Synthesisers *may* know the analysis to be conducted and users *may* have security clearance or agreement with the NSO, however the opposite may also be true.	High utility with possible varying levels of confidentiality.
Testing technology	The value of synthetic data is dependent on how complex the system is and how sophisticated the test data needs to be. Many methods to generate synthetic data may be too computationally heavy to make the effort worthwhile.	Medium utility and medium confidentiality.

ROBOT

Chapter 3:　Methods for creating synthetic data

There are many methods for generating synthetic data, and to determine which method to use, it is important to start by identifying the type of synthetic data that is required and within what context they will be used. Specifically, when creating a synthetic data set, the synthesiser (the individual making the synthetic data) needs to consider the desired analytical value to be preserved, as well as the acceptable level of disclosure risk, which will mainly depend on how accessible the synthetic data set that is generated will be (i.e., public release, restricted release, etc.).

With regard to the preservation of analytical value, the spectrum of available options is quite wide. Indeed, some projects only require specific pre-defined statistics and statistical conclusions to be preserved (such as the mean value of a given variable). At the other extreme, some projects require preservation of relationships between all of the variables to the maximum extent possible, without prior selection of specific statistics to preserve.

Recent developments in computing and software have expanded the range of methods available for generating different sorts of synthetic data. This chapter aims to provide an overview of such methods, and to establish recommendations on the most appropriate methods to use. The methods presented have been grouped into three categories:

- Sequential modelling;
- Simulated data; and
- Deep learning methods.

This chapter is focused on methods that have been used in practice by NSOs, although it should be noted that the field of synthetic data is always expanding, with further methods under development. While methods may be referred to in other parts of this guide, pertaining to such areas of further research or investigation, the goal of this chapter is rather to highlight the applicability of methods that have been explored by NSOs, and to outline the pros and the cons of these methods and the resulting synthetic data sets. In addition to methods, this section also presents some of the tools that can help to create synthetic data.

3.1　Sequential modelling

3.1.1　The Fully Conditional Specification (FCS) method

If the joint probability distribution of all of the variables in a data set is known, this not only reveals distributions for individual variables, but moreover the relationships between these variables. In such a scenario, simulating a synthetic data set from this information would in theory be straightforward to implement if sufficient computing resources were available.

In practice, however, the joint distribution will not be known, and must be estimated using modelling, though attempting to model the joint distribution in a single step is usually too difficult to do. Another option, therefore, is to decompose the (multidimensional) joint distribution into a series of conditional and *univariate* distributions that are easier to deal with, and this is the basis of the Fully Conditional Specification (FCS) method.

The FCS method was originally developed within a data imputation context (Van Buuren *et al.*, 2006), but given that data synthesis can be regarded as a massive imputation process on a data set, the FCS can also be used to create synthetic data. In the original imputation context, data are replaced because they are missing or invalid; in this new synthetic setting, valid data are being replaced to enhance the confidentiality protection. Instead of trying to explain all relationships between the variables that exist in the data set at once, the synthesiser proceeds step by step, by modelling and generating one variable at a time, each conditional upon the previous ones.

$$f_{X_1, X_2, \ldots, X_p} = f_{X_1} \times f_{X_2 | X_1} \times \ldots \times f_{X_p | X_1, X_2, \ldots, X_{p-1}} \qquad (1)$$

Data synthesis using the FCS can be implemented as a two-step process:

- First, the FCS is used to model the joint distribution, by using the original data set to estimate in turn each of the conditional distributions represented in the right-hand side of equation (1).
- The second step consists of generating synthetic values for each variable in turn, using the estimated model for the conditional distribution of that variable, using as input the synthetic values already produced for the previous variables (Drechsler, 2011).

Because the goal is to preserve the joint distribution as a whole, one could argue that the FCS aims to preserve all distributions and statistical conclusions rather than specific pre-identified summary statistics.

Two questions to consider when implementing this method are the order in which the variables in the data set are to be synthesised and generated, and the specific models to be used for each of the variables. However, there is no known standard procedure for selecting the order of the variables, and subject matter expertise may be important for informing such choices (for example, synthesising variables for age and level of education prior to synthesising income).

Models should be chosen carefully, considering the nature of the targeted variable, which allows the synthesiser some flexibility in how each variable is modelled. For example, some variables could be modelled using parametric models and others through non-parametric or mixed models. Classification and Regression Tree (CART) machine learning models are often a suitable choice for synthesising variables (Drechsler and Reiter, 2011). Indeed, they can more easily be implemented and adapted to data with irregular distributions than some other models (Reiter 2005). CART can notably capture non-linear relationships between variables which may not be properly considered with parametric modelling methods, which could be important in attempting to retain analytical value in the synthetic data set.

3.1.2 Pros, cons and considerations for the Fully Conditional Specification method

Table 2 summarises the pros and cons of the Fully Conditional Specification method.

Table 2   Pros and cons of Fully Conditional Specification

Pros	Cons
This method is relatively easy to understand and explain. Because the method estimates the joint distribution of the data set, this method aims to preserve all relationships between all variables, so it is not necessary to specify in advance which relationships between variables to preserve.	For skewed data (such as business or economic data), the presence of outliers remains a challenge in terms of disclosure or perceived disclosure control. With many variables the modelling process can become time-consuming.

3.1.3 Tools for FCS

The R package *synthpop* is a tool for generating synthetic data sets, implementing a range of different methods for doing so in a user-friendly manner. The majority of methods it incorporates are based on a full conditional specification (FCS), although methods for categorical data, based on preserving margins, are also available (Nowok *et al.*, 2015). For the FCS method, *synthpop* supports a wide range of ways to specify each of the conditional models, including both parametric and non-parametric models.

The package is designed to allow someone new to data synthesis to get started easily, by using the default options that are determined from the properties of the original data, while at the same time allowing flexibility for the more experienced user. It also includes tools for utility evaluation and for statistical disclosure control (SDC) of the synthetic output.[8]

3.1.4 **In practice:** Use of FCS at Statistics Canada

At present, Statistics Canada has released two public-use synthetic data sets that were generated using FCS, with high analytical value. Both of these synthetic data sets were created for use in hackathon-type activities, allowing their participants to draw statistical conclusions from the synthetic files that would be as close as possible to those which could be derived from the original data file, independently of the analysis performed.

The first of these events took place in 2018 as part of the 5th International Population Data Linkage Network conference held in Banff, Canada. To allow participants to undertake team-based analysis using a synthetic data set that mimicked original linked data, synthesis was performed using linkage of variables from the census and a mortality registry. The second of these events used a hackathon activity during the 2019 Canadian Cancer Research Conference, to explore relationships between cancer incidence, treatment and sociodemographic characteristics, based on data synthesised from a linkage of data from the census and administrative databases. Further details about the FCS method, its implementation and evaluation can be found in Sallier (2020).

3.1.5   The Information Preserving Statistical Obfuscation (IPSO) method

The goal of the Information Preserving Statistical Obfuscation (IPSO) method is to generate new synthetic data values, while preserving specific statistics and statistical conclusions (Cano and Torra, 2009).

The basic idea is that some of the variables in the data set are synthesised from other variables in that data set. For this purpose, the data set is partitioned into two subsets:

- The matrix Y, which is the subset of the original data file that is to be synthesised; and
- The matrix X, which is the subset of the original file that the model is based upon (containing explanatory variables).

The objective of this method is ultimately to create a synthetic version of Y, denoted Y', that can either be released on its own as a fully synthetic data set, or to be released together with X, as a partially synthetic data set (if the variables within X are non-confidential).

The first stage in this process is to make predictions for Y, denoted $\hat{Y}$, using a linear model relating Y to X as follows:

$$Y = \beta X + \varepsilon \quad (2)$$

where β is the set of model parameters, and ε are the set of residual errors between the actual values of Y and the model's predictions.

Assuming that the variables involved have multivariate normal probability distributions, then estimates for the parameters $\hat{\beta}$ can be made such that:

$$\hat{Y} = \hat{\beta} X \quad and \quad \varepsilon = Y - \hat{Y} \quad (3)$$

Then a normally distributed noise is added to $\hat{Y}$ to obtain the synthetic values Y'.

While there are different versions of the IPSO method, the main goal is to adjust the regression model, $Y = \beta X + \varepsilon$, in such a way that the synthetic values Y' provide the same (or very close) estimates of parameters (β) and variance-covariance matrix (Σ) as the original Y. Some versions of IPSO involve adding extra steps to force the equality $\hat{\beta}_{original} = \hat{\beta}_{synthetic}$, whereas some other versions are even stricter and force both equalities $\hat{\beta}_{original} = \hat{\beta}_{synthetic}$ and $\hat{\Sigma}_{original} = \hat{\Sigma}_{synthetic}$, so that if a user tried to fit the same model $Y' = \beta X + \varepsilon$, to the synthetic data set they would obtain the same estimates values for β and ε. This can be achieved by modifying the values of either Y' or X in such a way that equalities are respected. In the same vein, the synthesiser could decide in advance what specific parameters or summary statistic derived from the regression model they would like to preserve.

Hybrid methods can be obtained by completing existing methods with IPSO. Domingo-Ferrer and Gonzalez-Nicolas (2010) combined microaggregation with generation of synthetic data, running the IPSO procedure separately within each microaggregation cluster. The statistics preserved by ordinary IPSO are also preserved by this method. Using the method in Muralidhar and Sarathy (2008), it is possible at the variable level to select the degree of similarity to the original data. There is also random orthogonal matrix masking (Ting *et al.*, 2008) that controls the relationship with the original data via a single parameter.

By using one of these hybrid methods the problem of non-normal data can be reduced, which helps to fulfil the required modelling assumptions. More generally, Langsrud (2019) describes all of the above methods under a common framework, and within it develops improved algorithms and generalised methods.

3.1.6   Pros, cons and considerations

Table 3   Pros and cons of Information Preserving Statistical Obfuscation

Pros	Cons
Like the FCS, this method is fairly easy to understand and explain. With this method, it is possible to preserve some pre-identified parameters and summary statistics precisely, which can be used to allow a specific analysis performed on the synthetic data set to replicate results that would be obtained if the original data had been used. IPSO can be implemented as part of another method or process, to generate synthetic data sets. These hybrid methods may be used to alleviate the normal distribution assumption.	Normal distribution for all variables is a strong assumption that is seldom true.

3.1.7   Tools to apply IPSO

Examples and R packages for IPSO are as follows:

- Mu-Argus, Implementation of Domingo-Ferrer and Gonzalez-Nicolas (2010), https://github.com/sdcTools/muargus
- R package sdcMicro, An implementation of Ting *et al.* (2008) is included as a noise addition method, https://cran.r-project.org/package=sdcMicro
- R package RegSDC, Implementation of all methods described in Langsrud (2019), https://CRAN.R-project.org/package=RegSDC

3.2   Simulated data

3.2.1   From dummy files to more analytically advanced synthetic files

It is not always possible or practical to derive analytical results and conclusions using algebra, especially if the distribution of the data analysed is not known. Thus, a way to overcome this challenge is to use computer simulations that rely on a large number of repeated random sampling processes, to obtain numerical values and results. Monte-Carlo experiments for density estimation (L'Écuyer and Puchhammer, 2021) and Bootstrap procedures for variance estimation (Efron, 1979) are concrete examples where simulations are powerful tools for situations where algebra and equations become too complex to be solved.

Simulation processes can additionally be applied to the creation of confidential data sets, that can serve as synthetic data. For example, we could generate p independent vectors $X_1, X_2, \ldots, X_p$ of size N using a normal distribution generator process to obtain a synthetic data made of N synthetic units and p synthetic variables. In this example, the synthesiser could fix the values of the parameters to be used in the normal distribution generator process (i.e., the mean and the variance) without using any original data. In other words, synthetic data can be generated from 'scratch' without any disclosure risks. Here the analytical value would be considered to be null in the sense that no attributes of the original data would have been preserved in the synthetic data.

Such simulated data files are often referred to as "dummy" files within NSOs. However, it is important to realise that these dummy files can nonetheless be useful, depending on the users' needs. For example, if the goal is to test processes without any regards to the values or relationships between variables existing in the original data. This type of simulation process is easy to implement and can be used with other types of statistical distributions.

The synthesiser could also decide to use information from the original data in the generation process, to ensure that some of the analytical value is preserved. If we use our previous example, the synthesiser could have decided to generate the data is such a way that the parameters used to generate each of the p variables are estimated using original data for each of the variables. Thus, we would generate one synthetic variable for each original variable, using the estimated mean and variance observed for that original variable. In that case, the shape (or distribution) of the original variables might not be preserved (if the original variables do not follow a normal distribution) but the estimated means and variances for each of the synthetic variables would be the same in the original and synthetic data sets. Also, because the variables would have been generated separately, the relationships between them would not be preserved.

Simulation processes can be adapted to incorporate more or less of the information contained within the original microdata, so as to preserve more or less of its statistical properties according to its required use. For example, the Fleishman-Vale-Maurelli method derived from Fleishman (1978) and Vale and Maurelli (1983), is an approach that uses information from the original data to generate multivariate non-normal distributions with specific features preserved, such as intercorrelation between variables and marginal (univariate) means, variances, skewness and kurtosis. This method is well suited to capture correlations between continuous variables.

Options for capturing relationships between categorical variables include drawing values from the estimated multinomial equation, or classification and regression tree approaches, to create safe categorical variables.

3.2.2   Pros, cons and considerations

Table 4   Pros and cons of simulated data

Pros	Cons
Simulation processes are often easy to understand, and can create completely safe data when no information pertaining to the original data is used. However, for more advanced types of simulations, some analytical value can be preserved.	May not meet complex analytical needs, particularly if distributions or outputs need to match those from an original data set.

3.2.3 Tools to apply the method

In general, simulation processes can be programmed using a wide variety of software. For more complex types of simulation, the R package *semTools*[9] simulates microdata using the covariance matrix, skewness and kurtosis from the original sample data (Jorgensen *et al.*, 2019).

3.2.4 **In practice:** Australian Bureau of Statistics

In 2017 the Australian Bureau of Statistics (ABS) started using a serverless architecture (cloud), provided by the Amazon Web Server (AWS), for some of their projects. In order to explore emerging tools on their AWS, the ABS have generated completely safe synthetic data sets, using simulations, that can be sent out to the cloud to explore ML methods before obtaining approvals to access their data assets.

3.3 Deep learning

Deep learning is a subset of machine learning, and is a growing genre in the Data Science and Artificial Intelligence arenas. These methods are becoming more popular in the field of synthetic data because synthesisers are dealing more and more with large data sets. At the time of writing of this guide, Generative Adversarial Networks are a deep learning method that is used by NSOs to generate synthetic data.

With improvements in technology and computational capacity, implementation of machine learning processes has become easier and more accessible. Thus, it is natural that machine learning approaches have increasingly been employed to generate synthetic data sets. More specifically, the use of deep learning models has become appealing because of their capacity to develop powerful predictive models based on large data sets.

3.3.1 Generative Adversarial Networks (GAN)

The Generative Adversarial Network (GAN) (Goodfellow *et al.*, 2014) is a prominent generative model used to produce synthetic data. The model tries to learn the underlying structure of the original data by generating new data (more specifically, new samples) from the same statistical distribution as the original data, with two neural networks competing with each other in a game.[10] Because the GAN relies on neural networks, that means that the approach can be used to generate discrete, continuous or text synthetic data.

In a GAN there are two competing neural network models:

- One is called the generator and takes noise (or random values) as input and generates samples.

- The other model, the discriminator, receives samples from both the generator and the training data, and attempts to distinguish between the two sources.

---

9   https://CRAN.R-project.org/package=semTools

10   Because the theory and implementation of processes related to deep learning and neural networks
      can be technically challenging, we will mainly explain the overall concepts, as more information
      can be found in the references.

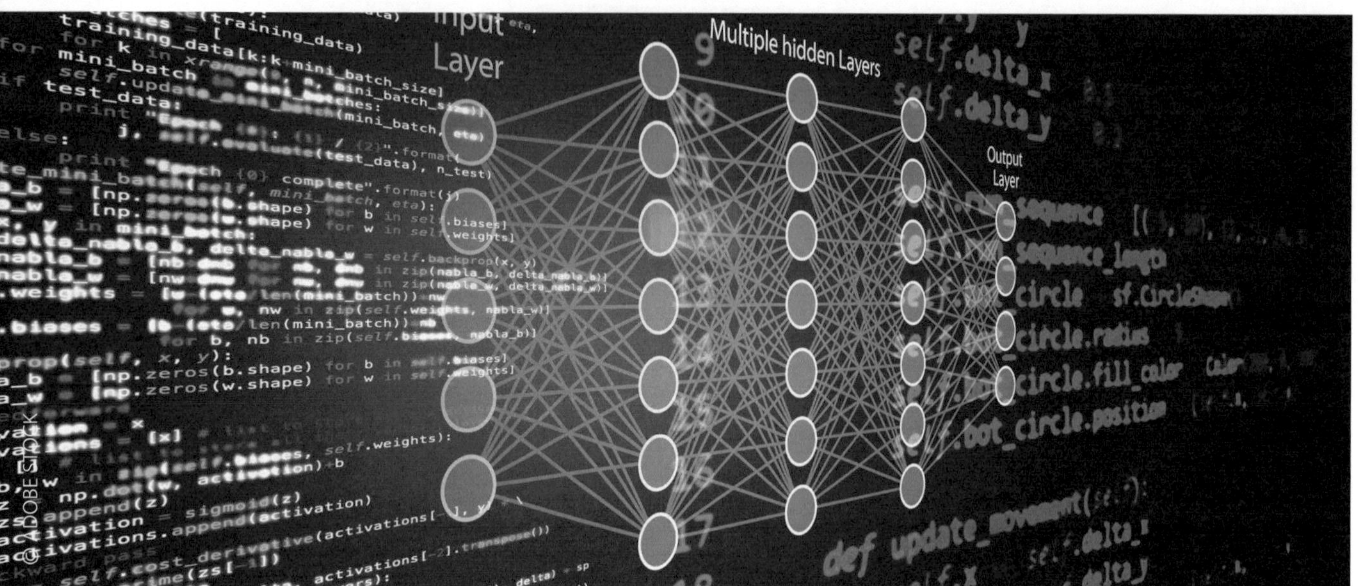

The discriminator serves a function similar to a binary classifier, that would take as input both real (or original) data as well as generated (or synthetic) data, and would compute a pseudo-probability value that would be compared to a fixed threshold value in order to classify the input from the generator as either generated or real.

As shown in Figure 4, the training process is an iterative one, during which the two networks play an ongoing game where the generator is learning to produce more realistic samples, while the discriminator is learning to get better at distinguishing generated data from real data. This interaction between the two networks is required for the success of GAN as they both learn at the expense of each other, eventually attaining an equilibrium.

Figure 4   Illustration of training of a GAN

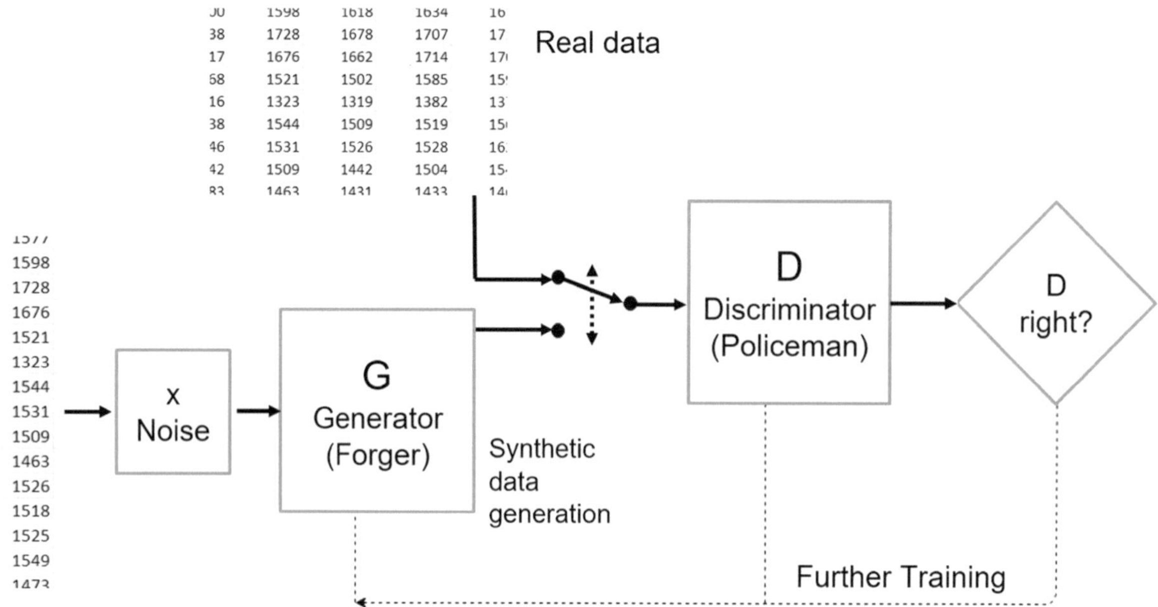

Source: Kaloskampis et al. (2020).

3.3.2   Pros, cons and considerations

Table 5   Pro and cons of GANs

Pros	Cons
GANs have been used in NSOs to generate continuous, discrete and textual data sets, while ensuring that the underlying distribution and patterns of the original data are preserved. Furthermore, recent research has been focused on the generation of free-text data which can be convenient in situations where models need to be developed to classify text data.	GANs can be seen as complex to understand, explain or implement where there is only a minimal knowledge of neural networks. There is often a criticism associated with neural networks as lacking in transparency. The method is time consuming and has a high demand for computational resources. GANs may suffer from mode collapse, and lack of diversity, although newer variations of the algorithm seem to remedy these issues. Modelling discrete data can be difficult for GAN models.

3.3.3   Tools to apply the method

There is no specific tool broadly available and used in practice in NSOs. However, Kaloskampis *et al.* (2020) provides detailed information on the method, and how to implement it in the context of NSOs. In addition, open-source tools are being developed for such models, for example the Synthetic Data Vault (https://sdv.dev/).

3.3.4   **In Practice:** Data Science Campus, Office for National Statistics (ONS)

The ONS Data Science Campus has explored the use of synthetic data to replace sensitive original data for testing their Census 2021 system. To communicate their work, Kaloskampis *et al.* (2020) published a study on generating synthetic data sets based on the U.S. Census Bureau's income data set.[11] This data set contains numerical and categorical variables, including socio-demographic information, and variables related to income (such as working status and income itself).

GANs were used in a binary classification model context, in order to generate new synthetic data. More specifically, the idea was to train a GAN algorithm to predict whether the income of an individual exceeds $50,000 per year based on some of the variables available in the original data set. In this example from the ONS, income is the target variable and the generator provides synthetic values at the end of the process.

---

11   available on the UCI repository: https://archive.ics.uci.edu/ml/datasets/adult

3.3.5 Other deep learning methods

This section presents very brief overviews of some other deep learning synthetic data generation methods that are gaining traction in the research and development communities, and which NSOs would benefit from being aware of.

Autoencoders are feed-forward deep neural networks, which are used to first compress and then decompress the original data. This is somewhat analogous to saving an image file at a lower resolution and then trying to reconstruct the higher resolution image from the lower resolution version.

The first part of the process is performed by a neural network of its own called the encoder, which restricts the amount of information that travels through the network using a convolution. Autoencoders use a second deep learning network called the decoder, which tries to reverse the effect of the encoder, by attempting to reconstruct the original input, with the reconstruction being synthetic data (Kaloskampis *et al.*, 2020). Figure 5 illustrates the architecture of an autoencoder.

Figure 5 Illustration of autoencoder architecture

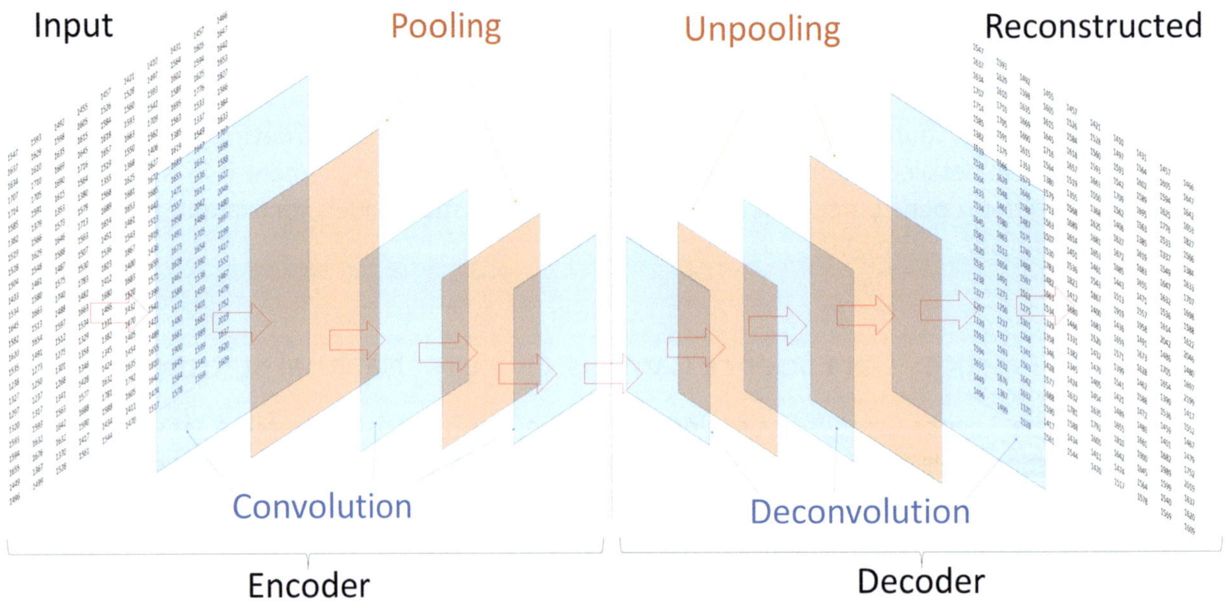

Source: Kaloskampis et al. (2020).

Autoregressive models are being explored to improve on some of the shortcomings of GANs models (Leduc and Grislain, 2021). Autoregressive models use a variant of a regression formula, that allows for the prediction of the next point of a sequence, based on previous observations of that sequence.

Other methods of note are Synthetic Minority Oversampling Technique (SMOTE) methods, which create synthetic data instances based on existing instances from the original data (Chawla *et al.*, 2002). Many of these deep learning methods are used to create differentially private synthetic data, which will be discussed in further detail in Chapter 4.

3.4 Methodological considerations

As synthetic data become more widely used in practical settings, those who synthesise it must consider dealing with issues such as variable types and sampling weights. This section presents possible solutions to some of the challenges that synthesisers may encounter.

3.4.1 Handling data types

The data holdings of statistical organizations consist of different types of data, such as numerical variables (e.g., age, and income) and categorical variables (such as marital status or occupation). Synthetic data are created by modelling the data, with the models used for this depending on the sort of variables in the data set, and their properties.

Some categorical variables may have an implicit ordering, such as salary bands, while others have no such ordering. Discrete variables (e.g., age), are often based on underlying continuous variables, but in official statistics they are expressed only to a limited accuracy (for age, in completed years or age groupings). Grouping of continuous variables into distinct intervals is often used to reduce disclosure risk.

When synthesis proceeds via conditional distributions (such as with the Fully Conditional Specification), the model used for each conditional distribution must be appropriate to the type of variable. For example, a logistic regression model can only be used for a categorical variable with two categories, and a log-normal distribution can only be used for a strictly positive numerical variable.

Some tools are available to help identify appropriate methods for the type of data, for example the *synthpop* package, to check if each variable is appropriate for the method being considered to model its conditional distribution. Transformations may be required for numerical data, to make the data more appropriate for the model. Some methods for modelling conditional distributions (e.g., CART) can be used for either numerical or categorical data.

Similarly, methods that model the whole distribution may require all variables to be numerical (e.g., Information Preserving Statistical Obfuscation) or all variables to be categorical (e.g., the differentially private histogram method, discussed in Chapter 4) while others (e.g., some implementations of GANs) can handle mixtures of different types of variables.

The synthesiser can create discrete variables by binning (grouping) continuous variables. This can be done during the synthesis, for example in the *synthpop* package, or via tools provided to use at a pre-processing stage. This allows methods for categorical data to be applied to data sets containing numerical data, although some of the information contained in the continuous variable is sacrificed.

Some variables are combinations of categorical and continuous variables. The most common example of this is a numerical variable with missing values. In this case, there are two data types for this variable. One is a 'missingness' indicator, and the other is the value of the numerical variable when it is not missing. Such data can be synthesised as part of a FCS, by first synthesising the missingness indicator, and then synthesising the non-missing values for cases where the synthetic missingness indicator is false. The *synthpop* package handles this automatically for missing values, and it can be customised for other types of mixed variable.

Both utility and disclosure risk, discussed later in this guide, depend crucially on the type of data being synthesised. For categorical data, a variable with many possible categories (and hence potentially small cell-entries) can pose a disclosure risk, but pooling groups of small categories can make the data less useful. Similar considerations apply to the precision with which numerical data are released. Particular values of continuous variables can pose a disclosure risk if the synthetic generation method reproduces unique values from the original data set. Smoothing or top-coding these values either during or after synthesis can mitigate the disclosure risk.

3.4.2   Data synthesis and survey features

Presently, most research related to data synthesis has focused on census data sets and administrative data. Lately, however, greater attention has been paid to exploring how to synthesise samples drawn from finite populations. A natural question which arises here is how to include survey design features in the data synthesis process, given that NSOs often use probabilistic sampling to collect survey data.

Most of the methods presented in this chapter work under the assumption that the original data cover the entire population of interest or are drawn via a non-informative sampling process from a finite population (such as simple random sampling). In other words, if the original data consists of sample from a finite population, the methods usually aim at synthesising the original data without regard for the finite population from which it was drawn. Therefore, statistical conclusions obtained via the synthetic file can only be comparable to the ones obtained in the original *sample* and not necessarily the original *population*, especially when the sampling process follows an informative design (Lavallée and Beaumont, 2015).

Thus, a way to address this would be to incorporate information on the sampling process within the data synthesis procedure, in order to obtain a synthetic dataset from which to estimate characteristics of the original *population*. This section aims at summarising some proposed strategies for generating synthetic survey weights or incorporating weights in the synthetic data generation process.

The sampling weight can be interpreted as the number of typical units in the surveyed population that each sampled unit represents. Often, auxiliary information (from a census or administrative file) about the survey population is used to compute survey weights. Weights may be further adjusted, for example to account for non-response, incorporating more auxiliary information (Statistics Canada, 2003). Then, estimates can be calculated using the survey weights.

Because of the use of auxiliary information, weights can often be a fingerprint of the information that went into producing them. Therefore, NSOs do not usually release survey weights, as this information could be used to identify specific units. Thus, it was a consensus among international experts that it would be unwise to release original sampling weights, even if the rest of the variables were to be synthesised, notably because if all units in a given sub-group of the population have the same distinct weight, then, even if the record is synthetic, that weight must have originated from data in that sub-group. In that case, if an attacker knows the mapping between weights and sub-groups, providing synthetic weights, or including weight in the data synthesis process, removes the exact fingerprint mapping between the weights and the original sampled units.

Alternatively, weights can inherently be very distinct, which is more commonly an issue with asymmetric data, such as business data, where large businesses may have weights that look unlike any other business. In this case, even synthetic weights pose a risk of outlier re-identification if the synthetic weights resemble the original weights, and either differential privacy could be a solution, or this might motivate the inclusion of weights directly in the data synthesis process such that no weights (even synthetic) would be directly released to users. In either case, it is important to realise that the disclosure risks of releasing synthetic weights can be estimated the same way as the disclosure risk of any other variable is estimated.

Then two types of strategies were raised:

1. Generate and provide synthetic weights to users.

    a. Treat the weights as a variable to be synthesised among all the others.

    b. Synthesise the design variables and recalculate synthetic sampling weights based on them and the original sampling design.

2. Use weighted models to approximate distributions from the original *population, and generate values from them (in this case no weights would be provided to users)*.

3.4.2.1 Generate and provide synthetic weights to users

Some users might want to have synthetic weights made available with the rest of the synthetic variables, especially if the goal is to explore the synthetic data set while waiting for access to the original data set if that is also expected to include weights.

3.4.2.1.a Treat the weights as a variable to be synthesised among all the others

The idea was to consider survey weights as any other variables to be synthesised. One of the main concerns here is to ensure that the synthesised weights need to be coherent with the synthesised design variables as well as all the other variables.

3.4.2.1.b Synthesise the design variables and recalculate synthetic sampling weights based on them and the original sampling design

Here, the goal would be to synthesise all variables involved in the sampling design, and to recalculate synthetic weights in the same way that they were calculated at the design stage of the survey. However, that covers mainly the calculation of *basic* sampling weights. Usually, in practice, basic weights are then being processed to take into account calibration, non-response and even smoothing on some occasions. For example, for calibration, it could be possible to adjust the synthetic weights to ensure that some weighted totals of the synthetic file correspond to known totals from the original population. Also, for non-response adjustment, a solution could be to re-weight for total non-response with the synthetic weights. In that case, it would be important to synthesise total non-response indicator variables in the synthesis process and then use those to re-weight synthetic weights with traditional methods.

In general, providing synthetic weights requires a lot of attention to be paid to the original survey design, and where those weights are coming from. In addition to that, it is important to realise that it also limits the selection of synthesis approaches that can be used. Indeed, recalculating synthetic sampling weights might be more complicated with more complex survey designs. Thus, it is important to understand users' needs, to ensure that the effort put in the synthesis process truly matches the analytical needs of the requirement.

3.4.2.2 Use weighted models to approximate distributions from the original *population*, and generate values from them

The pseudo likelihood method is suggested in this case. This method can be considered to be an example of an advanced simulation process that generates data of high analytical value. The idea is to preserve univariates statistics and relationships between variables from the original *population* as much as possible. In this case no weights would be provided to users.

3.4.3 Pseudo likelihood

The pseudo likelihood method generates synthetic populations, by incorporating survey weights into the models using the pseudo likelihood approach (Kim *et al.*, 2020). The idea is to estimate the distribution of the finite population. Once the finite population density is estimated, the synthesiser can generate fully synthetic populations, by drawing values repeatedly from it. This notably requires derivation of the full conditional distributions of the Markov Chain Monte Carlo (MCMC) algorithm for posterior inference, by using the pseudo likelihood function. A full presentation of the method is beyond the scope of this guide; further details can be found in (Kim *et al.*, 2020).

Pros, cons and considerations

Table 6   Pros and cons of Pseudo Likelihood Method

Pros	Cons
When generating synthetic populations, the sampling process is already accounted for; thus, the uncertainty introduced by the sampling process is also accounted for. Providing synthetic populations can be better than providing synthetic samples and can be more convenient (no need for original survey weights and no need to estimate sampling variance).	There are potential challenges with the choice of prior distribution for the MCMC algorithm.

Tools to apply the method

There is no known tool *per se* to apply the method. However, section 2.1 and 2.2 of Kim *et al.* (2020) provides detailed information on the method and how to implement it.

3.5 Tips to get started

This section provides recommendations on synthetic data generation methods, depending on the particular use case and requirements for the synthetic data produced. Table 7 connects the recommended methods with the use cases presented in Chapter 2. Figure 6 illustrates a decision tree to help practitioners identify the method most suitable for their project.

Table 7   Method recommendations by use case

Methods		Use Cases			
		Public synthetic microdata release / testing analysis	Education	Testing Technology	Comments
Sequential Modelling	Fully Conditional Specification	Recommended	Can be used, though if analyses conducted and statistical conclusions are pre-determined it might be too time-consuming in comparison to other methods.	Can be used, but might be too advanced in comparison to the real analytical need.	This method aims, in theory, at preserving relationships between variables from the original data. Disclosure risk and analytical value need to be evaluated according to the release process.
	IPSO	Recommended if the analyses are all related to linear regression, otherwise not recommended.	Can be used, depending on the context: Recommended if the analyses are all related to linear regressions, otherwise not recommended.	Can be used, but might be too advanced in comparison to the real analytical need.	This method preserves results and statistics specifically related to linear regressions. Disclosure risk and analytical value need to be evaluated according to the release process.

Table 7   Method recommendations by use case (continued)

Methods		Use Cases			
		Public synthetic microdata release / testing analysis	Education	Testing Technology	Comments
Simulated Data	Dummy files	Not recommended	Can be used if training does not require analytical value in the data.	Recommended	This method does not preserve any analytical value from the original data, but is easy and quick to implement. Data is totally safe.
	Analytically advanced simulated data	Recommended if analyses conducted are related to the pre-identified results that needed to be preserved in the synthesis process. Otherwise, not recommended.	Recommended if analyses conducted are related to the pre-identified results that needed to be preserved in the synthesis process. Otherwise, not recommended.	Can be used, but might be too advanced in comparison to the real analytical need.	This method only preserves pre-identified results and statistics. Disclosure risk and analytical value need to be evaluated according to the release process.
	Pseudo Likelihood	Strongly recommended if user wants to estimate statistics from the original finite population. However not recommended if user is expecting synthetic weights.	Can be used, though if analyses conducted and statistical conclusions are pre-determined it might be too time-consuming in comparison to other methods.	Can be used, but might be too advanced in comparison to the real analytical need.	This method aims, in theory, at preserving relationships between variables from the original population. Disclosure risk and analytical value need to be evaluated according to the release process.
Deep Learning	Generative Adversarial Network	Recommended especially in presence of text or unstructured data.	Can be used, though if analyses conducted and statistical conclusions are pre-determined it might be too time-consuming in comparison to other methods.	Can be used, but might be too advanced in comparison to the real analytical need.	This method aims, in theory, at preserving relationships between variables from the original data. The only method that handles unstructured and text data. Disclosure risk and analytical value need to be evaluated according to the release process.

Figure 6   Synthetic data generation method decision tree

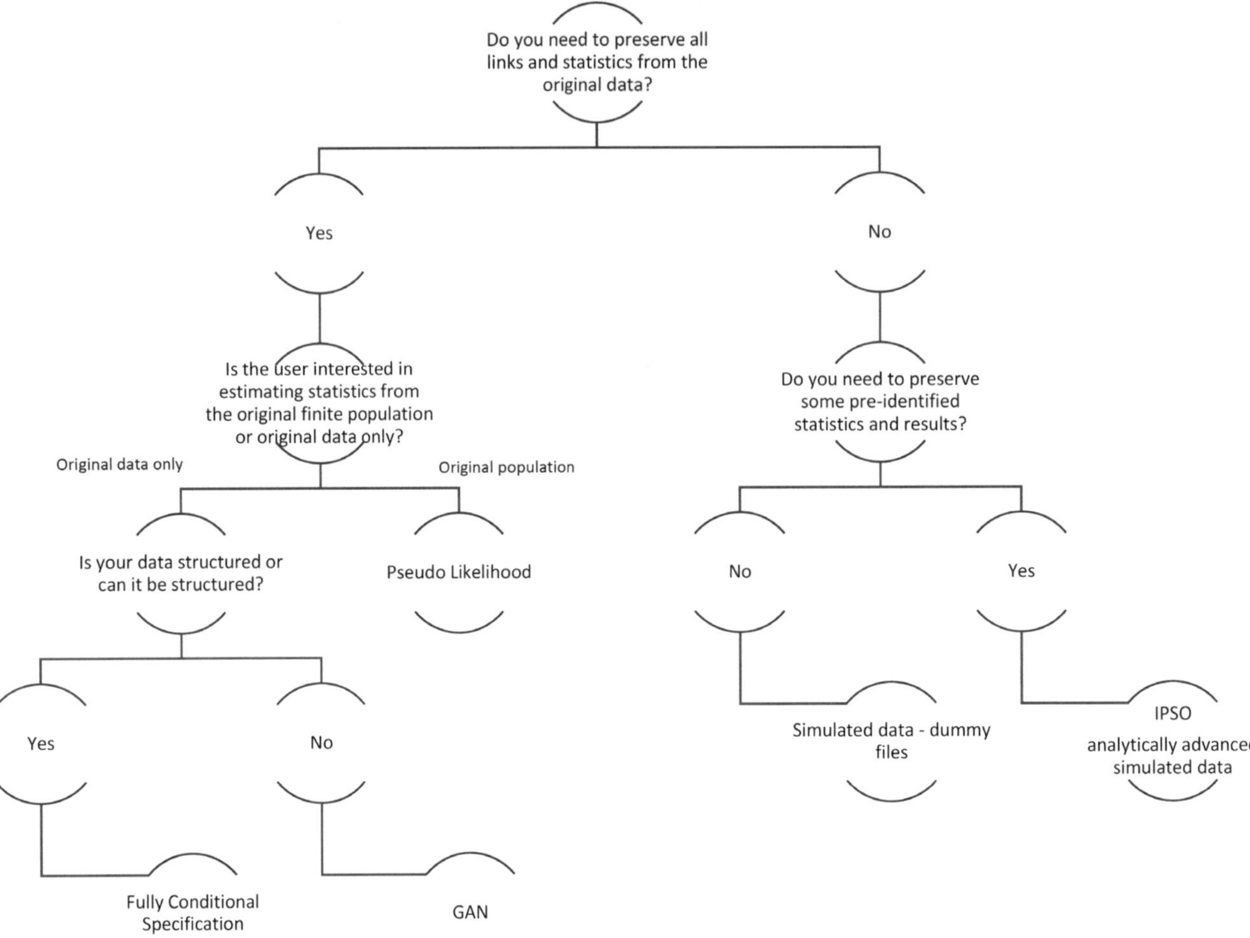

Chapter 4:   Disclosure considerations for synthetic data

According to the Organization of Economic Cooperation and Development (OECD, 2003), disclosure relates to the inappropriate release of data or attribute information of an individual or an organization. Disclosure risk is the possibility of disclosure occurring, and disclosure control refers to methods that can be used to reduce it. The purpose of this chapter is to present the disclosure control options available to NSOs and their synthesisers.

When data are altered to protect against disclosure, either by creating synthetic data or by other methods, the discussion is often in terms of the trade-off between disclosure risk and utility. This is often presented graphically as shown in Figure 7. For synthetic data, as we discuss in the next section, utility is a measure of the closeness of results from the synthetic data compared to the original data. Data where all the records had identical values might have a utility of zero, and a disclosure risk of zero, while the original data will have a utility of 100%. Disclosure risk for synthetic data is at a maximum when the original data are unaltered. The ideal position on this graph, at the top left corner, with perfect utility and no disclosure risk, is never attainable because altering the data to protect against disclosure risk will always change data values.

Figure 7   The trade-off between utility and disclosure risk

There has been much less work on developing disclosure risk measures for synthetic data than for measures of data utility, with the exception of the disclosure risk measure, *epsilon (ε)*, that is defined for differentially private (DP) synthetic data (to be further discussed later in this chapter). An overview of DP methods for synthetic data is given below.

Epsilon is not a disclosure measure that is computed after the synthetic data have been created, but a parameter that is set by the person creating the synthetic data. *Epsilon* determines the variance of the noise that is added to the original data with small values of *epsilon* corresponding to a large amount of noise and small disclosure risk.

Although no record in a (fully) synthetic data file corresponds to a real person or household, there is concern that attribute and identification disclosure risk could still be present. Identification disclosure could be present if unique observations found in the population are present in the synthetic data (Drechsler and Reiter, 2009) and attribute disclosure would occur if the value of the original data for such an individual could be determined, perhaps wrongly, from the synthetic data. These situations could result in loss of reputation for the data holders and put at risk respondents' willingness to participate in surveys, census or provide their information by other means. Therefore, NSOs may still decide to use additional disclosure controls in addition to synthetic data.

NSOs should choose whether or not to implement additional disclosure controls on their synthetic data, as well as any specific privacy preserving techniques, based on their own legislative and operational frameworks. Common privacy preserving techniques and other statistical disclosure control (SDC) measures can be applied to both synthetic as well as real data. Examples include top-coding or bottom coding of extreme values and smoothing or rounding of numerical data as well as indistinguishability-based techniques such as k-anonymity, ℓ-Diversity, t-closeness that are frequently used by NSOs. We include introductions to these methods here, and towards the end of this chapter we outline methods for disclosure control that have been developed by Statistics New Zealand.

4.1   Measures of disclosure risk for synthetic data

There are two main types of disclosure: identity disclosure and attribute disclosure. Identity disclosure occurs when a record in the released data is recognised as matching to an individual for whom the attacker knows values of the released data from another source. This definition is only relevant to partially synthetic data since the records in fully synthetic data do not have one-to-one matches to the original data. A small subset of variables in the data, including the synthesised values, are used to make the linkage and once the linkage is successful, the attacker can disclose all other non-synthesised attributes for the individual. Reiter and Mitra (2009) have proposed methods for evaluating disclosure risks for partially synthetic data. They assume that an attacker has access to multiple versions of the synthetic data and uses a sophisticated approach to evaluating this risk that will not be suitable for NSOs.

Attribute disclosure from either fully or partially synthetic data can occur without identity disclosure. The scenario we envisage is that an attacker observes an individual in the synthetic data that appears to be a unique match to a known individual. An attacker may use synthetic data to obtain information for an individual about one or more of their synthesised values. Here we need to distinguish between information that is available from the model that has generated the synthetic data and information that is specific to an individual in the released data. As an example of the former, the synthesis model for the relationship between occupations and income might find that a high proportion of merchant bankers were in the highest income brand. Then an attacker with the synthetic data, wishing to determine the income band for a merchant banker, would be able to do so with high probability.

4.1.1 Attribute disclosure measures

Approaches developed to address attribute disclosure from researchers in the context of synthetic data have used the idea of *replicated-uniques*, also sometimes termed *apparent matches*. A replicated unique is a record in the synthetic data with a unique combination of values identified from all variables or from a selected subset and where this combination is also a unique record in the original data. These are termed *key* variables or *quasi-identifiers* that represent information that an attacker might possess about an individual. An attacker might then assume that this was the real individual with these characteristics and find the values of other items in the person's record. Of course, such values may be wrong, but publicity associated with an apparent data breach could be detrimental to the NSO. Thus, NSOs can take steps to delete *replicated-uniques* or to minimise the proportion of such records present in synthetic data. Nowok *et al.* (2017), have investigated this method and its impact on utility for synthetic data. As expected, the proportion of *replicated-uniques* is greatest for small data sets with pseudo identifiers taking many unique values. In the examples investigated, the removal of *replicated-uniques* had only a very small influence on data utility.

Replicated-uniques or *apparent matches* can also be used as the basis of a disclosure risk measure for synthetic data. The *apparent matches* in the synthetic data can be aligned with the records in the original data. This was illustrated with data synthesised from the year 2018 of the American Community Survey (ACS) that was part of the *HLG-MOS Synthetic Data Challenge 2022*, an accompanying sprint to evaluate this guide (Bhagat *et al.*, 2022).[12] The original data contained 143,371 records and it was synthesised with the default settings of the *synthpop* package. There were 1,826 records in the synthetic data with unique combinations of the quasi-identifiers, PUMA area, age, and sex. Only 671 of these records, or 0.47%, were also unique records in the original data, giving a 0.47% rate of *replicated-uniques*.

To mimic what an attacker might do, we can then match each of the *replicated-uniques* back to the original data; a sample of 9 such records is shown in Table 8, where we also show 3 other variables that might be of interest to an attacker, MARST (marital status), EDUC (education level) and INCTOT (total income). For example, we can count the proportion of categorical variables where the synthetic and original data agree. In the sample shown 4 of the 9 records agree on marital status and 3 on education but only one on both. The percentages for all 671 *replicated-uniques* in the ACS data were 55% for MARST, 37% for education, and 16% for both. For INCTOT we could calculate the percentage of synthetic incomes that were within 20% of the original data for the 9 records in Table 8 there is only one such. This was true for 16% of the 671 records.

---

12   The *HLG-MOS Synthetic Data Challenge 2022 was held in January 2022 in order* to evaluate the recommendations of this guide in practical applications. A total of 17 teams participated, representing NSOs, industry and academia, from various countries. Participants were tasked to synthesise data with different combinations of data sets, synthesis methods and utility and disclosure measures. The results and experiences from this challenge contributed to content within this guide. The results and knowledge gained from the challenge is archived on https://pages.nist.gov/HLG-MOS_Synthetic_Data_Test_Drive/index.html (Bhagat *et al.*, 2022)

Table 8   Sample of 9 randomly chosen records from the 671 replicated uniques in the synthesised 2018 ACS data.

Quasi-identifiers			Other variables					
			Synthetic data			Original data		
PUMA	Age	Sex	MARST	EDUC	INCTOT	MARST	EDUC	INCTOT
17-3532	72	Male	With spouse	7	20,000	With spouse	8	139,600
17-3527	64	Male	With spouse	8	80,000	Single never married	11	37,000
39-5403	37	Male	No spouse	10	44,000	No spouse	6	32,000
17-3521	76	Female	Widow	6	51,700	With spouse	6	23,000
17-3308	85	Male	With spouse	11	74,000	With spouse	7	22,500
39-4107	86	Female	Widow	6	31,680	Divorced	6	17,100
17-3523	88	Female	Single never married	2	19,000	Widow	5	18,050
17-3528	86	Female	Widow	6	16,700	Widow	6	79,800
17-3309	84	Male	With spouse	6	75,700	Single never married	8	26,000

Source: HLG-MOS Synthetic Data Challenge 2022 (Bhagat et al., 2022)

Summary measures of disclosure risk could include the proportion of *replicated-uniques* as well as the proportion of non-key variables that were correctly identified from *apparent matches*. Depending on the results of such analyses the user might decide to delete all *replicated-uniques* based on quasi-identifiers from the data. If they are left in the released data, the summary measures would give some measure of the likelihood of an attacker getting a correct answer to an attempt to identify an individual's attribute. This approach has not been fully developed at present, and further work is needed to find out if it could be of practical use.

4.1.2   Identity disclosure measure

Identity disclosure can be more difficult to measure, particularly with a fully synthetic data set where no record is real. A method that has been utilised by one NSO, at the writing of this guide, to assess identity disclosure is called *rates related to database reconstruction*. These rates refer to the percentage of matches one gets from reconstructed data to the original data, with the intent to determine how easily the original data can be generated with the information available to the public. As an example, a subset of the Summary File 1 (SF1) of the 2010 Census of the United States of America was taken to review. The SF1 contains data from the questions asked of citizens of the United States of America and about every American housing unit. The SF1 has the following data protections: the housing units in the file were swapped at some unpublished rate; the group quarters in the file were protected using synthetic data techniques and the tables themselves are an information reduction. The data for the variables, age, sex, race, Hispanic/Latino ethnicity, and census block (geography) variables were reconstructed using a subset of the SF1 through a system of equations that when solved, converted to microdata. These records were then matched with data from commercial holdings as well as the Census Edited File (CEF) itself. In the commercial holdings example, the U.S. Census Bureau had a specific series of commercial data holdings and this experiment was to determine if identities could be inferred via linking data sets. The CEF experiment was conducted as a worst-case analysis, serving as a proxy for an attacker with data holdings on par with the CEF in terms of quality. More generally, the exercise could be conducted with any additional data sets that contain identifiers, like name and address, and a subset of variables to match with the subset under review (for example age and sex). In this Census example, after combining the reconstructed with the commercial data, the five variables plus a unique identifier are linked exactly to the same variables in the CEF. There are two main rates to consider: putative (suspected match) and confirmation linkage, or confirmed match.

The matching process works as follows:

1. Reconstructed data says there is 1 male white Hispanic person aged 52 in a block.
2. The commercial data shows that the only 52-year-old male in that block is Bob at X address (putative match).
3. Attach white/Hispanic to the putative match.
4. Look for Bob/X address/age 52/male/white/Hispanic in the CEF.
5. Find that record in the CEF (confirmed match).

This exercise can be scaled to rather large data sets, as demonstrated by the exercise conducted by the U.S. Census Bureau. Table 9 summarises the putative, confirmed, and precision rates of the census exercise. Precision is defined as the confirmed matches divided by the putative matches. The precision variable in Table 9 shows how often we are right when we think we are.

This method was implemented to assess re-identification via matching all the variables. However, it is important to keep in mind that if the commercial (or other) data holdings used for re-identification purposes contain other variables on the file that can be attributed to an individual. Thus, we need to consider attribution disclosure as presented in the precious section. In summary, rates related to database reconstruction pertain to identity disclosure but strongly imply the potential for attribute disclosure.

Table 9   Disclosure Risk Assessment of Population Uniques by Block Population Size

Block Population Bin	Putative Re-identifications (Source: Commercial Data)	Confirmed Re-identifications (Source: Commercial Data)	Precision (Source: Commercial Data)	Putative Re-identifications (Source: CEF)	Confirmed Re-identifications (Source: CEF)	Precision (Source: CEF)
Total	137,709,807	52,038,366	37.79%	238,175,305	178,958,726	75.14%
0						
1-9	1,921,418	1,387,962	72.24%	4,220,571	4,093,151	96.98%
10-49	25,148,298	13,481,700	53.61%	47,352,910	43,415,168	91.68%
50-99	30,567,157	12,781,790	41.82%	51,846,547	42,515,756	82.00%
100-249	38,306,957	13,225,998	34.53%	63,258,561	45,807,270	72.41%
250-499	21,789,931	6,408,814	29.41%	35,454,412	22,902,054	64.60%
500-999	13,803,283	3,460,118	25.07%	23,280,718	13,514,134	58.05%
1000+	6,172,763	1,291,984	20.93%	12,761,586	6,711,193	52.59%

Source: Abdow, J. (2021b)

4.2   Privacy-preserving techniques

Privacy preserving techniques can be applied on real data or synthetic data. NSOs may choose to apply privacy preserving techniques in addition to synthetic data based on their national context, legislative and operational frameworks. Of increasing interest is the differential privacy framework and how it can be applied in conjunction with synthetic data.

4.2.1   Differential Privacy

Differential Privacy (DP) was introduced in 2006 in computer science by Cynthia Dwork, Frank McSherry, Kobbi Nissim and Adam Smith. It is only recently that DP has made its way to official statistics and DP-compliant methods are playing an increasingly important role in today's digital world. While DP's terminology is well established making use of computer science terms such as databases, mechanisms and queries in this document we adopt a more statistical language to convey the same notions.

Contrary to popular belief, DP is neither a method nor an algorithm but a definition supporting a mathematical disclosure-control framework. Thus, despite its name, the intended use of DP in official statistics is to prevent disclosure when *releasing statistical information* rather than to address privacy concerns when *gathering personal data* from individuals. As such, DP presents itself as an alternative to the traditional disclosure control framework which has been used by NSOs for decades now.

The two main vintages of DP are: the one-parameter ε-DP and the weaker two-parameter (ε-δ) DP. In this chapter, we focus on the more stringent ε-DP, which is the form people usually have in mind when referring to DP. Where applicable, DP can be used by a data custodian to provide explicit and mathematically provable disclosure protection guarantees *when releasing statistical information derived from personal data*. DP is the first disclosure control framework explicitly stating the type and degree of protection it is offering.

What is a DP_ε-compliant Method?

Any method applied to a data set D for the purpose of releasing statistical information - either in the form of tables or data sets (including those of a synthetic kind) - is at risk of disclosing some of the personal information D contains. Such a method M is called (ε-) differentially private or DP_ε-compliant when it meets DP's mathematical requirements which impose restrictions on the *type* of disclosure that may occur.

Furthermore, the privacy parameter ε - whose value in practice is set by the data custodian - determines the *degree* of disclosure protection a DP_ε-compliant method M is offering by means of the upper limit it imposes on the amount of person-level information M *might* be disclosing. More explicitly, the data custodian controls through ε the (average) amount of suitable random noise used by M to produce its outputs, which incidentally also impacts their utility: the degrees of protection and utility are closely linked. Indeed, the larger (smaller) the value set to ε the less (more) noisy M's outputs become and, generally speaking, the greater (lesser) the disclosure risk they pose and the greater (lesser) their utility.[13]

---

13   ε is often referred to as the "privacy loss parameter", with small values leading to less loss of privacy, and larger values leading to greater losses of privacy.

Why the name, Differential Privacy?

To better understand DP we need to state the problem it has been introduced to solve. As a preamble suppose you were told that John's height is 10 centimetres (or 4 inches) more than the national average for men. Now should you learn from a data set of national heights, say, that the national average is 1m78 (or 5 feet 10 inches) then you would correctly conclude that John is 1m88 (or 6 feet 2 inches) tall.

Note that, and this is key, John's height may have become known to you *without him ever having participated* in the data collection of the national heights data set. In the situation where indeed, John did not contribute his information to the data set, then its custodian can hardly prevent disclosure from happening since it is directly due to *external* factors such as what people (like you) already know about John. After all, while it is certainly John's prerogative to not contribute information to this data set, he cannot prevent *others* from contributing their information which may bring the data set to play an accessory role in a disclosure event involving him.

By thinking of examples such as this one, the authors of DP posited that disclosure protection guarantees cannot be absolute in nature: simply put, a method ought not to be held responsible for a disclosure event that *could have occurred without* the concerned individual ever having contributed to the data set itself. Instead, they proposed assessing a method's role in a disclosure event differentially *by comparing* what it can reveal about an individual *when their information is present* in the data set *to when their information is not included*.

In effect, this differential assessment limits M's role in a disclosure event to what it can reveal *as a result* of the participation (or absence thereof) of the concerned individual to the data set. Hence, if M's outputs were about the same regardless of whether John had participated in the data set or not, then M would hardly be at risk of disclosing something specific about John since his contributions have little influence on M's outputs. This will become clearer later when examples of a DP_ε-compliant method are presented.

What are exactly the mathematical requirements underlying Differential Privacy?

The concerns just expressed over the role any one piece of information might play in determining M's output led to DP's mathematical requirements. But first, those concerns must be rephrased in the language of data sets that M understands. To that end, consider *two* data sets that are identical *except* for a single piece of information one has and the other not; in DP's parlance these are called adjacent data sets. Then, *for a given* output of the method M *and a given pair* of adjacent data sets D and D', determine *how more likely* is this output value to occur when M is applied to D instead of D'.

The relative likelihood assessment just described involves a *specific* combination of an output *and* a pair of adjacent input data sets. DP-compliance requires doing the same, but this time for an *arbitrary* combination of an output value and a pair of adjacent data sets. More specifically, M is DP_ε-compliant *if* an output is no more likely to occur than the limit set by the data custodian through the privacy parameter ε when M is applied to D rather than to D', *irrespective of which one output and which one pair* of adjacent data sets D and D' are considered.

This is DP's way of sizing up the influence *any one individual in the data set* might have on *any one of* M's outputs, as their personal information may be the only thing separating D from D'. This is not to say that the upper limit associated with a DP_ε-compliant method holds for *all* outputs *and all* pairs *simultaneously*; it rather means that it is holding for *any one arbitrary* output-pair combination, as opposed to holding just for one *cherry-picked* combination.

Furthermore, the upper limit calculated as part of DP's likelihood assessment has to hold *even if* the user already knows everything about the data set's contents *except* for just one individual's contribution *and* knows the inner working of *M*. Thus, DP has what can be called a no-secrecy policy: the validity of its protection guarantee does not rely on users having little knowledge of both the data set and the disclosure control method used, an assumption usually made for the protection offered to be most effective. On the contrary, DP's likelihood assessment is to be performed by assuming users have an understanding of *D* and *M* comparable to that of the data custodian. We will discuss this further below when examples of a DP-compliant method are presented. DP's no-secrecy policy is a major selling point and explains to a large extent practitioners' interest in DP as a disclosure control framework.

What does it mean to have formal disclosure protection guarantees?

The explicit and mathematically provable protection guarantees DP offers characterise this framework. While such formal guarantees are a novelty in a disclosure control context, they are a familiar occurrence in probability sampling: statistical bias and variance are examples of formal *quality* guarantees. For both probability sampling and DP, formal guarantees stem from the mathematical requirements underlying each framework.

In contrast, it is usually *not* possible to formally assign bias and variance measures to estimates derived from nonprobability samples, which were *en vogue* in the 1930s before probability sampling was introduced and have recently been making a comeback. While a carefully designed nonprobability sample might allow for sound inferences to be made, it has traditionally been lacking the mathematical underpinnings needed to support formal quality guarantees such as bias and variance.

What does a DP_ε-compliant method look like?

Just like a probability sample *looks* the same as a nonprobability sample (both are just subsets of the population), a DP_ε-compliant method *looks* no different than any other method. The difference rather lies in how the samples and methods are *conceived*: both a probability sample and a DP_ε-compliant method result from a process designed to meet well-defined mathematical requirements from which they each inherit their formal guarantees. While some of the dissemination methods exploited under the traditional disclosure control framework already are DP_ε-compliant, most are not (although some can be modified to become differentially private).

Applications: two examples of a DP_ε-compliant method

In this section two examples of a DP_ε-compliant method are given for illustrative purposes: the first releases a numerical value and the other a data set. In each case, the DP_ε-compliant method is derived from a non-compliant one to showcase its distinctive features.

First, consider the situation in which a data custodian is looking to release a count extracted from *D* after rounding it to the *nearest 5*, say. Hence, raw counts of 14 and 12 would get released as 15 and 10, respectively. This method is *not* differentially private. Indeed, a key theorem of DP (i.e., *a verifiable consequence of DP's mathematical requirements*) states that no deterministic method can be differentially private. And rounding to the *nearest* 5 is such a method since, for instance, it *always* returns 15 when the raw count is 14 (and *always* returns 10 when the raw count is 12). Note that increasing the rounding base to 10 or 100 would not change the outcome: it is the deterministic 'nearest-to' feature of the rounding that poses the problem here, not the actual base value.

It is instructive to establish non-compliance directly, that is without appealing to DP's theorem. Consider a data set D containing 13 yeses, which gets rounded *up* to 15, along with its adjacent D' obtained from D by dropping one of those thirteen records. The ensuing count of 12 yeses for D' would get rounded *down* to 10, allowing one to conclude *with certainty* that the dropped record was a 'yes'. The same argument, but expressed this time in DP's parlance, shows that the output value of 15 occurs with *probability one* when the rounding is applied to D (with its 13 yeses) but with *probability zero* when the rounding is applied to D' (with its 12 yeses). This can be paraphrased as saying that M's output of 15 is "infinitely" more likely to occur from D rather than from its neighbour D', which is something DP does not allow to happen as "infinity" is larger than any pre-set limit.

From this, we conclude that a method must have a random component to be differentially private. However, this is not a sufficient condition. For example, rounding a raw count to the multiple of 5 directly above or below it based on the flip of a coin (e.g., rounding 14 to either 10 or 15, each with probability ½) has a random component but is *not* differentially private. Indeed, consider D leading to a count of interest of 15. Since it is already a multiple of 5, it would get released as 15. Now consider D' such that its count is 15-1=14. Under the current random rounding scheme, this count has equal chances of getting released as either 10 or 15. Because the output value 10 is *possible* for D' but *impossible* for D, their relative occurrence likelihood is "infinite" which again is something DP does not allow to happen.

In this situation, a DP_ε-compliant method is Laplace's mechanism, which is obtained by adding noise generated from Laplace's *continuous* distribution to the raw count. For example, 15.199385… and 13.836519… are just two of the *countless* possibilities for Laplace's output to the same input value of 14. (Later we discuss a way of using Laplace's mechanism to release *whole* numbers, which is what users would naturally expect to get as released counts.)

Laplace's mechanism[14] was the first example ever given of a DP_ε-compliant method and it possibly remains the simplest one to this day. The variance of the Laplace distribution determines the (average) amount of noise that gets generated. Not surprisingly then, the privacy parameter ε of Laplace's mechanism is directly tied to the variance of the underlying Laplace distribution, which drives the degree of protection (and utility) conferred by the data custodian to Laplace's outputs.

Considering users' dislike of noise-adding methods, a welcome consequence of DP's no-secrecy policy is that a data custodian *can safely divulge* the variance of Laplace's distribution used (or equivalently, the value set to the privacy parameter e). This is a departure from traditional disclosure control practices which call for such information to be kept secret in order for the protection offered to be most effective. Thus, under DP users are given the means to assess the statistical significance of the conclusions they are drawing by factoring in the (average) amount of noise that has gone into the outputs analysed. It is important to note that in practice, accounting for the noise in the output is extremely difficult. To date, there is a knowledge gap between NSOs and users on interpreting noisy results. Inexperienced users are often not comfortable looking at negative counts. Some NSOs, for example, the case of the Australian Bureau of Statistics, have found that there are significant investments in educating users so that they can be comfortable with non-additive tables.

---

14   While it is tempting to use the better-known Gaussian or normal distribution instead of Laplace's, this strategy does not quite lead to a DP_ε-compliant method since it only meets the requirements of the weaker two-parameter $(\varepsilon-\delta)$ form of DP.

For the second example of a DP_ε-compliant method, suppose a Yes-No question is administered to individuals using a Randomised Response (RR) method as a means of reducing the response bias due to the sensitive nature of the information gathered – see for instance Section 12.5 in Lohr (1999) for a short discussion of RR methodology. A RR method introduces plausible deniability[15] by altering any given response with probability p (which is known to the data custodian) before recording it in a data set. Also, a RR method is to keep no trace of the reported answers nor of which ones were altered.

It can be shown that knowing p allows one to draw meaningful statistical conclusions about the true proportion of yeses from the *recorded* values alone. However, in the traditional disclosure control setting only the data custodian would be allowed to know p. In contrast, in the case of a DP_ε-compliant RR method, the value for p could be safely divulged in accordance with DP's no-secrecy policy allowing users to make valid inferences as well.

We can see such a RR method as producing partially synthetic data, albeit of a very limited analytical kind. Indeed, not only are cross-relations among variables not captured by the method discussed here, but there is also legitimate ground to question its efficacy in addressing the initial bias concerns.

For the sake of further illustrating how DP-compliance works, suppose the data custodian felt[16] that only yeses were sensitive and therefore needed to be protected. To this end, a RR method is used whereby only some of the yeses are randomly altered: all reported noes are directly recorded in the data set.

While economical in the amount of noise it uses, this RR method is *not* DP-compliant. To see why, consider a data set as output of this method in which a 'yes' has been recorded for a certain individual. The custodian might argue that without knowing how the method works, a user cannot say *for certain* whether this value points to a reported 'no' instead of a reported 'yes'. However, this line of argumentation does not comply with DP's no-secrecy policy which requires assuming the user *does know* how the method works. And a user knowing that a reported 'no' cannot possibly be recorded as a 'yes' in the output data set by this RR method would conclude *with certainty* – which is something DP does not allow to happen – that the individual had to have answered 'yes' to the question. In this situation, DP-compliance requires randomly altering some of the noes as well *to complete* the masking of the yeses undertaken.

Is a DP-compliant method always better than a non-compliant one?

The short answer is no, not always. Just like an ill-designed probability sample can lead to nonsensical conclusions (as exemplified by the famous tale of Basu's elephants [17]), an ill-devised differentially private method can do a poor job of preserving the personal information used as input. If the method is not well designed or the data is too difficult, the added noise can overwhelm the input, rendering the released outputs all but useless.

---

15   The notion that respondents can deny that the recorded answers are truly those they initially expressed.

16   As argued above, deciding which information is to be protected ought to be a matter of the agreement passed between parties and not a matter of opinion as is the case here.

17   First told in Basu (1971), numerous other accounts of the story are widely available on the Web.

Previously we discussed the basic approach for adding Laplacian noise to counts of individuals. Data synthesis is often performed by combining these privatised counts to parameterise or train a model which can generate new records in the original schema. Simple histogram models, probabilistic graphical models (PGM)s and generative adversarial networks (GAN) have all been used for this (Bowen and Snoke, 2021). These methods will be elaborated on later in this chapter, however intuitively, for high utility we want the added noise to be small in comparison to the original count values, so the relative shift between the input and output counts is not large. More accurate counts lead to more realistic data models. But, due to differential privacy composition, the more counts we take over a given individual, the more noise we need to add to maintain the same level of privacy.

This means that challenging cases occur if a given data schema has a large number of variables (which require many counts for the model to capture), or the data has few records (meaning many counts are small), or if the variables have many possible values (meaning the data is spread sparsely across different options and many counts are small or zero). These conditions can pose significant problems for utility that researchers are actively working to overcome (Bowen and Snoke, 2021). However, when a data set has a large number of records, a small number of variables, and does not spread the data out too sparsely, existing DP synthetic data generators can produce very high-quality data with robustly protected privacy. Pre-processing data can often improve performance by eliminating variables and reducing granularity on variables with large numbers of possible values.

In addition, DP methods can fall short on protecting privacy as well. If the privacy parameter is excessively large and the method does minimal additional processing to the data, the provided protection can deteriorate to the point of being meaningless.

Also, a commonly held belief has DP providing *unfailing* protection against disclosure of personal information, which it does not. For instance, DP does not prevent the effective disclosure protection from decreasing when multiple outputs involving the same individual are released. For example, the random noise used in Laplace's mechanism tends to cancel out when several of its outputs are averaged out to form a single result. On the bright side, not only does DP's Compositional Theorem warn data custodians of the compounded privacy loss incurred by making repeated use of a DP_ε-compliant method to release statistical information, it also *quantifies* that loss. It states that the composite privacy loss ε is the sum of the privacy losses for each output. This has given rise in practice to the notion of privacy budget which is used to closely monitor the situation: once the cumulative privacy loss incurred from a series of releases made from a data set reaches the budgeted value set by the data custodian, access to the data set is closed.

Some Implementation Considerations

A successful implementation of any new methodology to a specific context promises to be a challenge, as various practical constraints will bring issues not directly addressed by the theory. And looking to release differentially-private synthetic data is not an exception. However, attention to the following details will certainly help with the implementation of DP in such a context.

With respect to implementing data synthesis within a national statistical agency, Sallier and Girard (2018) recommend decomposing the required tasks into pre-processing, synthesis, and post-processing steps. In their experience (which did not involve DP considerations), a successful implementation of data synthesis hinges on making informed, forward-looking decisions at the pre-processing stage of the project. Indeed, while it is tempting to rush straight into synthesising a data set, it is important to first reflect on the balance to be struck between utility and confidentiality. By avoiding making decisions based on utility alone, one prevents putting too much strain on synthesis to protect confidentiality, whether this is attempted under DP or not.

To illustrate, consider the following example. A data set D contains information about families, including the number of siblings each contains. Clearly, the participation of a family with ten siblings will have a greater impact on the *total number of siblings* than on the *total count of families*. Indeed, the latter statistic inherently offers greater anonymity since a family with ten siblings only contributes a value of 1 to the tally, as all other families do. Pre-processing involves deciding here whether information about family size remains in the data set or not; and if it does, then cutting off its tail by resorting to an open-ended category such as "more than 5 siblings" would reduce the burden put onto synthesis.

When designing a DP_ε-compliant method to meet practical needs, two important consequences of DP's mathematical requirements may prove very useful: the Compositional Theorem and the Post-Processing Theorem. We saw previously that the Compositional Theorem warns data custodians against the compounded privacy loss incurred by repeated use of a DP_ε-compliant method on a data set. But the same theorem can also be used to manufacture a nontrivial DP_ε-compliant method by putting together two existing and simpler $DP_{\varepsilon/2}$-compliant methods. Thus, a complex DP_ε-compliant method need not be created from scratch, but it can rather be built using available or easier-to-design DP-compliant pieces.

We alluded to the Post-Processing Theorem before (although not by name) when evoking a way of releasing Laplace's outputs as integers; loosely put, it states that any transformation of the output of a DP_ε-compliant method is itself DP_ε-compliant *provided it is operating independently of the input data set*. Thus, to get an integer out of Laplace's mechanism one simply needs to round Laplace's output to the *nearest* integer. In the same vein, the Post-Processing Theorem supports the practice of setting a negative Laplace's output to 0 prior to being released without upsetting its DP-compliance status. However, if one were to round Laplace's output to the nearest integer *only when* the raw count is smaller than 15, say, then this would violate the theorem's premise. Indeed, since Laplace's output itself does not reveal whether the original raw count was smaller than 15 or not, this rounding rule can only be implemented by first revisiting the data set. But then because Laplace's mechanism is no longer the only way statistical information is getting released from the data set, its previously-earned DP-compliance status gets revoked. (This is not a statement of non-compliance per se – it merely implies that DP-compliance of the combined Laplace-with-conditional-rounding method must be examined anew.)

But wait, why is it rounding Laplace's output to the *nearest* integer leads here to DP-compliance when we previously established that rounding a raw count to the *nearest* 5 (or 10, 100, etc.) is not itself a DP_ε-compliant method? The apparent contradiction is resolved by paying close attention to *what* gets rounded in each case. While DP says that rounding *a raw count* to the nearest integer is not effective at preventing disclosure of personal information from happening, rounding *Laplace's output* to the nearest integer poses no issue when it is solely done for practical reasons. Indeed, in this case, Laplace is the one method responsible for protecting the personal data that has gone into the output, not the subsequent rounding performed. The concern here rather becomes whether the extra rounding performed can *undo* the protection already provided by Laplace's mechanism. DP says this will not happen *as long as* Laplace's mechanism remains the only way statistical information gets released from the data set.

Differentially Private Data Synthesis

How can differential privacy be applied in the context of synthetic data? If an organization wishes to apply differential privacy to their synthetic data, there are two broad categories of methods that have been explored at length: marginal based methods and GANs.

A broad family of marginal based methods range in simplicity with the histogram method to the more complex Probabilistic Graphical Models (PGM). This family of methods takes counts of the number of records that have a given combination of feature values, noise (based on the differential privacy framework) is then added to those feature values and then are used to generate synthetic data.

The *histogram method* is useful when there are few features (variables) so that one can take the direct count of the number of records that fall into each possible combination of features. The histogram method transforms continuous variables into discrete variables through binning the data. Noise (following the differential privacy framework) is then applied to the frequencies of the constructed bins. The final synthetic data is then obtained by sampling from the noisy bins proportionally to the noisy counts (Wasserman and Zhou, 2010). It proceeds as follows:

1. Group all continuous variables in the data into categories or bins, where the optimal choice for the total number of bins is $M(n, r)$ is given by: $M(n, r) = n^{r/(2+r)}$ where n is the total sample size and r is the number of continuous variables

2. Produce a cross-tabulation of all combinations (the histogram)

3. Add Laplace noise to each cell of the table

4. Create a multidimensional sampling grid, where each point falls in the centre of one bin. Each point in the grid will be assigned a weight equal to the noisy count of the bin to which the point belongs.

5. A weighted sample will be drawn from the grid using the weights assigned in step 4.

6. To smooth the selected sample, apply noise from a uniform distribution such that each selected point will move randomly from the centre of the bin to somewhere inside the bin. These new re-spaced sample represent the synthetic data.

Figure 8 provides an example of a simplified histogram generation method with a toy data set with two variables. For illustrative purposes, three bins are shown instead of the optimal number (as determined by $n^{1/(2+r)}$). The addition of Laplace noise (step 2) to the true counts in each bin (shown as a 3D histogram in panel B) yields noisy counts (panel C). The noisy counts are used to create a two-dimensional sampling grid (step 3; panel D) where the weights for each bin are used when sampling (step 4) the number of points that become the synthetic data (panel E). The new points can be thought of as centred in the grid but are shown spread out so that the reader can see how many there are in each bin. The points are then spaced out within a bin through the addition of uniform random noise (step 5) to redistribute them across the range of each bin for the continuous variable (panel F). Panel G compares the newly generated synthetic data to the original.

Figure 8 Illustration of a simplified histogram generation method with a toy data set with two variables

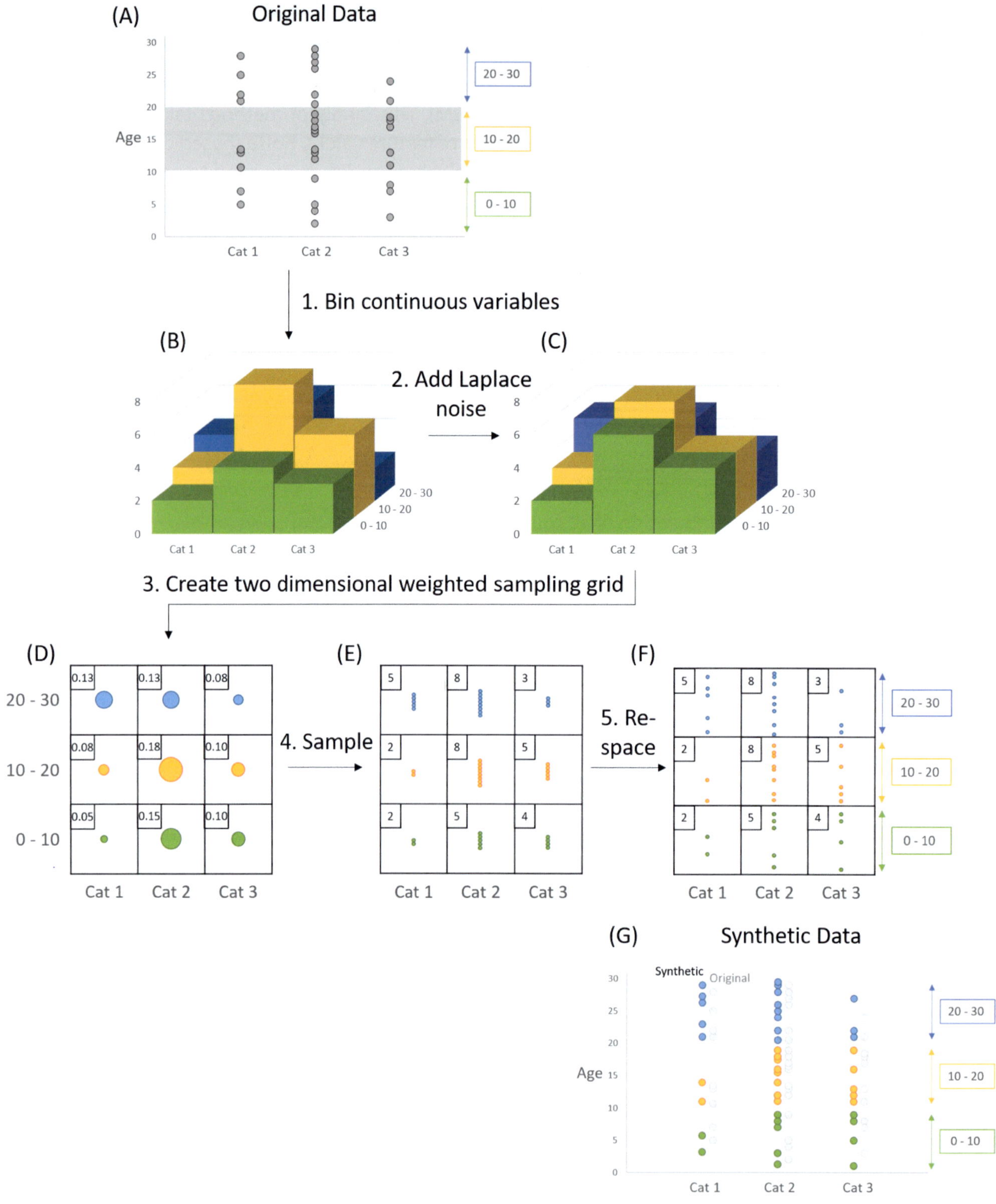

The addition of Laplace noise to the histogram bins, following Dwork *et al.* (2006), makes the histogram method differentially private. However, this perturbation can lead to changes in the distribution of the data. When there are bins with zero (or low) counts, adding Laplace noise will yield some bin counts above zero and some below. Those below get truncated to zero resulting in a noisy distribution that is different from the original. Such changes in the histogram distribution are most likely to occur when the original data is sparse or if the bin size is set such that there are many bins with zero (or low) counts. Despite changes in the individual bin counts, the histogram method can still generate exactly the same number of synthetic records as the original data using a well-designed sampling procedure.

However, if you have a lot of features, it can be impossible with the histogram method to capture the distribution as counts for every possible combination for all features at once. To make this more manageable, *marginal methods* are used to break up the features to be more manageable. Marginal methods use the marginal distributions of a subset of random variables to determine the probability distribution of variables contained in the original data (Ridgeway *et al.*, 2021).

Adding noise can increase the difficulty to get from the counts to the synthetic data as noisy counts may not match up to what was present in the original data. Given a key feature of synthetic data is its replication of distributions and relationships of the original data, the mismatch in counts between the original and noisy data is problematic on the utility front. PGMs can be used to get from pairwise (2-marginal) probability distributions to a fully generated set of synthetic records. PGM uses graph-structured interpretable models to record patterns of variable correlations. These graphs are constructed automatically from the original data. The graphs are then manipulated using reasoning algorithms to create synthetic data (Ridgeway *et al.*, 2021).

Within the context of marginal and PGM methods, differential privacy strictly protects all information associated with any single individual, which means that synthetic data generators which satisfy differential privacy cannot directly deidentify individual records or samples from real individual records. Instead, they need to take as input noisy *aggregate* statistics about the population as a whole, where the added noise has been calibrated to satisfy the DP guarantee. The composition theorem shows that the ε for the synthetic data is then the sum of the εs for all of the statistics used to define the model fit. Marginal methods used most often to create synthetic data define the model from a set of margins, each of which has noise added to make it DP. These marginal methods create synthetic data using noisy 'marginals' or co-occurrence counts on features. Given a collection of marginals, noise is added and so these noisy counts no longer reflect any single consistent set of records. The noisy number of records with a certain set of attributes might not be resolvable into any single set of counts about whether those records have the exact match of real attributes. This could then be rendered into a set of synthetic records with all of the features in the initial schema. Marginal methods can use probabilistic graphical models (Bayesian models, Markov models), constraint satisfaction ("solving" for a synthetic data set that comes closest to matching all of the noisy counts), or even iteratively weighting or rescaling public data to mirror the noisy counts on the private data.

Additional computation is required in the case of marginal methods to make the margins consistent with each other. This is justified by the post-processing theorem (Bowen and Snoke, 2021) which states that once a DP result has been released, any subsequent transformation of these outputs are also DP. An important limitation of these methods is that any exploration of the data to define the model to use for the synthesis must contribute to the privacy budget. Since DP requires additional noise with each additional query, there is pressure to capture the distribution of the data with as few queries as necessary, and this means there are a variety of creative techniques for getting from this optimised set of noisy marginal counts back to complete synthetic records. If the model is selected from another similar data set, say from a previous year's version of the data release, then this information can be used freely without contributing to the privacy budget.

The GANs approach to creating differentially private synthetic data involves iteratively training a deep learning model for the synthetic data with a differentially private mechanism, which involves adding noise in each iteration. As a reminder from Chapter 3, GANs mainly consist of a generator and a discriminator. In each iteration, the discriminator trains a binary classifier from the original data and the synthetic data produced by the generator and attempts to distinguish between original and synthetic data. In the training process, the model minimises an empirical loss function to improve the data fit. Non-privacy preserving GANs use optimisation algorithms such as the statistical gradient descent (SGD) for the minimisation. Differentially private GANs use a noisy SGD to incorporate ε into the discriminator during the training process (Abadi *et al.*, 2016). The ε of the differentially private mechanism is accumulated from each training iteration. Adding to Figure 4 from Chapter 3, Figure 9 illustrates where ε will come into play in the GANs synthetic generation process.

Figure 9   Illustration of training of differentially private GAN based on Kaloskampis *et al.* (2020).

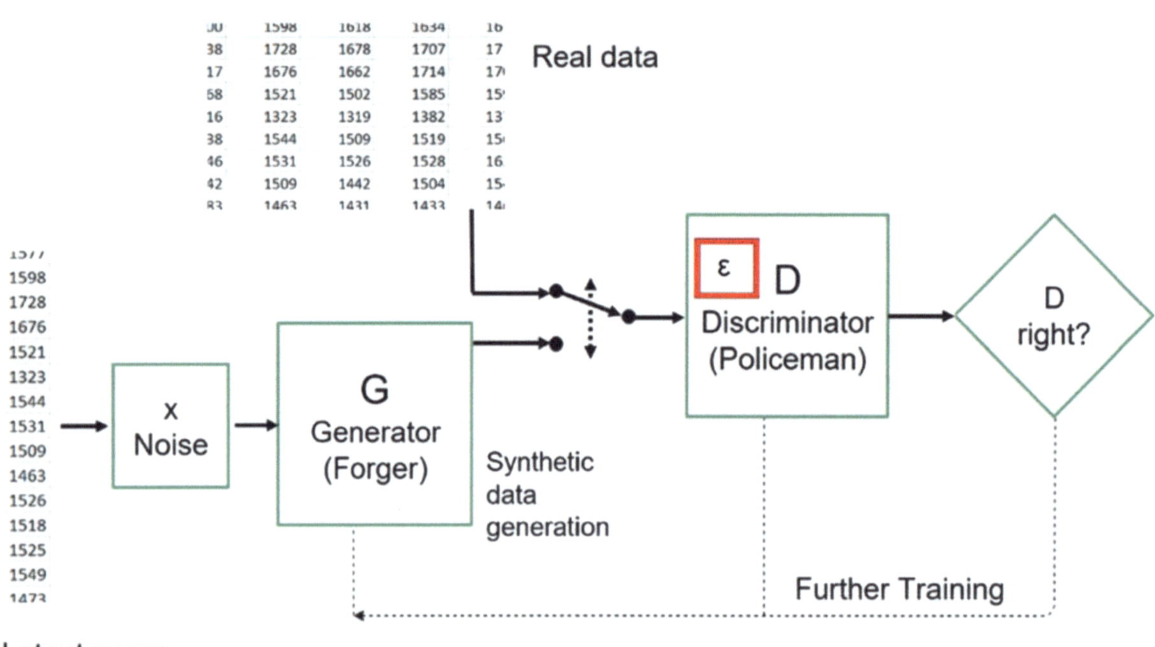

Source: Based on Kaloskampis et al. (2020).

In these methods, each iteration of the training process uses a noisy objective function over the input data to improve the model fit, and the noisy objective function is one form of noisy aggregate statistic of the input data. In these iterative approaches, the ε from these procedures is the sum of those from each step. Unlike non-DP iterative procedures, the number of steps must be fixed in advance so as to maintain the DP property.

To summarise this section, Table 10 presents the pros and cons of differentially private synthetic data methods.

Table 10   Pros and cons of differentially private synthetic data generation methods

Pros	Cons
These methods create synthetic data that adheres to the differential privacy framework. Adhering to the differential privacy framework can be useful as it provides a level of privacy assurance, making it easier to make decisions and communicate with end-users on the level of privacy.	In most evaluations of utility, these differentially private methods do not perform as well as their non-differentially private counterparts (source). Many of these methods can computationally be quite costly.

Tools to apply these methods

As the field of differentially private synthetic data evolves so too do the tools to generate it. Open-source tools for PGMs are being more widely used. Examples of such tooling include Private-pgm (McKenna *et al.*, 2019) and DataSynthesizer (Data, Responsibly 2021).

Why are DP-compliant methods not more widely used in official statistics?

DP is the first framework capable of addressing formal disclosure protection guarantees. However, the underlying mathematical requirements are often difficult to satisfy in practice and many practitioners actually find them too stringent to begin with. More specifically, they see the disclosure scenario underlying DP as too severe: a user will simply not know just about everything that is contained in the data set as DP requires the custodian to assume through its mathematical requirements. As a result, they claim, DP's requirements unduly undermine the utility that can ever be attained. DP_ε-compliant methods may in fact not provide the required utility under different NSO's requirements. The DP framework was developed under cryptography literature and the main objective is to protect privacy. Protecting privacy is an important consideration for NSOs, but protection needs to be balanced with providing useful statistical information to inform decision-making. Deployment of the DP-compliant methods will be context dependent to maximise NSOs ability to provide useful statistical information.

While DP's requirements are indeed strong, there is more to them than getting data custodians to assume the existence of some super-user. First and foremost, DP's requirements exist to ensure that the guarantees offered by custodians do not depend on what they have assumed users *do not know*. For instance, when designing their disclosure control strategies data custodians may very well underestimate the extent of the information already available to users prior to making their own releases. And if they do, then the protection guarantees offered will be weakened by what users actually do know.

Even when data custodians do have a good sense of what users already know in terms of a-priori information, their protection strategy could be compromised by a critical piece of information that comes to light after they make their releases. Thus, DP's requirements exist to also provide protection against unforeseen risk factors rather than just against some conjured omniscient user.

Without knowing what the future holds, the beginnings of DP are reminiscent of those of probability sampling back in the first half of the 20th Century. As Olkin (1987) points out from a conversation held with survey pioneer Morris Hansen, survey practitioners initially found it difficult to comply with the requirements of probability sampling. The first probability sample designs often were too rudimentary to meet realistic survey needs and there were many issues (e.g., how to devise unbiased estimators, how to assess their variance and how to deal with out-of-scope units, domain estimation, and nonresponse) for which adequate answers only came later. Thus, we can expect a body of best practices to emerge from the lessons learned as DP is applied to specific contexts e.g., Hawes (2020).

Currently, complex survey features such as clustering and hierarchical structures (e.g., family-related information on records pertaining to individuals) make it challenging to find DP_ε-compliant methods of any practical use in a survey context. Also, not only can it be difficult in certain circumstances to devise a useful non-trivial DP-compliant method but formally establishing its compliancy can prove to be a daunting task.

One can expect more flexible mathematical requirements to be proposed in the years to come allowing for formal disclosure protection guarantees to be used in a wider array of practical situations than what is currently possible under DP. For example, while we know of ways to implement a DP-compliant method to release a count, it is not quite clear how to proceed for the total of a non-dichotomous variable such as income. The wider the range of values a variable can take, the larger DP's upper limit tends to become in order for it to hold for all possible outputs which can render a DP-compliant method all but useless in practice. Also, further guidance will be needed on how best to handle survey-specific features such as sample design information including survey weights. The flurry of DP-related papers published these last few years is a testament to the efforts being deployed to put forward formal disclosure protection guarantees beyond what the pioneering work of Cynthia Dwork, Frank McSherry, Kobbi Nissim and Adam Smith has provided already through DP.

Finally, even without seeking to achieve compliance, NSOs can still benefit from DP by reviewing their current disclosure control practices in light of its principles to identify and close gaps in the protection offered. For example, we know from a theorem of DP that a method needs to have a random component to be DP_ε-compliant. This suggests that an NSO rounding a count to the nearest pre-set base value would gain from using some *random* rounding method instead, thereby making its practices presumably less prone to disclosure than they were before.

4.2.2   K-Anonymity

K-Anonymity is one of the most well-known privacy preserving techniques and is an example of one of the traditional disclosure methods mentioned in the section on differential privacy. The K-anonymity method gives K-level of anonymity to data, which means the information for each record contained in the release cannot be distinguished from at least k-1 other records whose information are in the data. Records can be associated with each other by certain identifying attributes, such as age, gender, and location, in the case of say census or medical records. Breaches or attacks can occur when these attributes, called quasi-identifiers, can be linked with external data to identify unique records in the population (Machanavajjhala *et al.*, 2007). K-Anonymity model distorts quasi-identifier values so that no record is uniquely identifiable from a group of k records. The Parameter k indicates the degree of anonymity (Sweeney 2002).

There are two main methods to achieve k-anonymity: suppression and generalisation. Suppression is a method of ensuring privacy by selectively hiding confidential information before disclosure. Cell suppression is the main method of data suppression. Under this methodology all sensitive cells are suppressed from publication, sometimes including non-sensitive cells as complementary suppression to obscure the values of the sensitive cells.

Generalisation coarsens an attribute to a more general value (Lefevre *et al.*, 2005). This creates groups of individuals that share the same generalised attribute value. There are two types of generalisation that can be done: full domain vs local generalisation. Full domain generalises all values of an attribute to the same level. Local generalisation generalises values of an attribute to different levels.

For an example a k-anonymity, refer to the example of records with sensitive medical information in Table 11.

Table 12 contains data on 12 individuals with a record of their medical condition. The medical condition of each individual is considered sensitive, meaning that an adversary must not be allowed to discover its value. Whereas neighbourhood (as defined by the first three digits of a postal code called Forward Sortation Area or FSA), age, and occupation are considered non-sensitive. In this example, the quasi-identifier is the combination of Postal Code, Age and Occupation attributes.

Table 11 Example of records with sensitive medical record information

	Non-sensitive			Sensitive
	FSA	Age	Occupation	Medical Condition
1	A1A	27	Teacher	Heart Disease
2	A1B	28	Electrician	Diabetes
3	A1C	29	Teacher	Cancer
4	A1D	24	Doctor	Cancer
5	C3E	35	Teacher	Cancer
6	C3E	37	Electrician	Diabetes
7	C3R	40	Doctor	Heart Disease
8	C3O	40	Teacher	Diabetes
9	C2R	50	Electrician	Cancer
10	C4M	48	Doctor	Heart Disease
11	C8S	49	Doctor	Heart Disease
12	C8Z	50	Teacher	Cancer

Table 11 contains an example of k-anonymity achieved by utilising generalisation and suppression (denoted by *) techniques. For instance, we aim to achieve 4-anonymity. This means that the values for FSA, age, and occupation of the individual records should be generalised in such a way that we can form equivalent classes with at least four records. The quasi-identifiers of these records should be indistinguishable from each other.

Table 12   4-anonymous version of Table 11

	Non-sensitive			Sensitive
	FSA	Age	Occupation	Medical Condition
1	A1*	2*	*	Heart Disease
2	A1*	2*	*	Diabetes
3	A1*	2*	*	Cancer
4	A1*	2*	*	Cancer
5	C3*	≤40	*	Cancer
6	C3*	≤40	*	Diabetes
7	C3*	≤40	*	Heart Disease
8	C3*	≤40	*	Diabetes
9	C**	≤50	*	Cancer
10	C**	≤50	*	Heart Disease
11	C**	≤50	*	Heart Disease
12	C**	≤50	*	Cancer

The main strengths of k-anonymity are its simplicity and potential to protect against re-identification attacks. Re-identification attacks can either happen through linking records in data sets or through multiple queries to the same database to obtain relational inferences.

However, k-anonymity assumes that each record in a data set represents a unique individual. If this is not the case, an equivalence class of K records does not necessarily link to K individuals with k-anonymity (Mendes *et al.*, 2017).

Both attribute and identification disclosure are still at risk. There are three types of attacks that can be used against k-anonymity: unsorted matching attack, complementary release attack, and temporal attack (Sweeney 2002).

Unsorted matching attacks is based on the order in which the groups appear in the released tables. For instance, if the released tables have the same order of the generalised groups, then a direct matching of groups across table positions can reveal sensitive information. This can be prevented by randomly sorting the order of the groups for the released tables.

Complementary release attack is based on finding quasi-identifiable attributes that are a subset of the attributes in complementary data sets that have been released. By obtaining quasi-identifiers through multiple data sets that are published, re-identification attack is possible. This can be prevented by doing a thorough inspection of external information.

Temporal attacks are based on temporal inference. Since a k-anonymity solution of a data set at time t has no requirement to respect the k-anonymity solution of the same data set at time $t+1$, then joining the two k-anonymity solution data sets can release sensitive information. This is prevented by doing a thorough inspection of external information and adjust for potential quasi-identifiers as well.

4.2.3 ℓ-Diversity

As mentioned above, there is still the potential for an attacker to identify information, even if k-anonymity is met. ℓ-Diversity is an extension of K-anonymity. It requires every equivalence class to abide by the ℓ-Diversity principle. An equivalence class is ℓ-diverse if at least ℓ "well-represented" values exist for the sensitive attributes (Machanavajjhala *et al.*, 2007).

Continuing the example from the previous section, Table 13 reflects a 3-diverse version of Table 11.

Table 13   3-diverse table of medical condition data

	Non-sensitive			Sensitive
	Postal Code	Age	Occupation	Medical Condition
1	A1*	≤30	*	Heart Disease
2	A1*	≤30	*	Diabetes
3	A1*	≤30	*	Cancer
4	A1*	≤30	*	Cancer
5	C**	≥30	*	Cancer
6	C**	≥30	*	Diabetes
7	C**	≥30	*	Heart Disease
8	C**	≥30	*	Cancer
9	C**	≤50	*	Heart Disease
10	C**	≤50	*	Heart Disease
11	C**	≤50	*	Cancer
12	C**	≤50	*	Diabetes

Distinct 3-diversity means that each equivalence class should contain at least three distinct values for the sensitive variable "medical condition": Heart Disease, Cancer, Diabetes. Thus, there should at least be three records in each equivalence class.

ℓ-Diversity increases privacy protection compared to K-anonymity (Li *et al.*, 2007). It protects against attribute disclosure and addresses the vulnerability to unsorted matching attacks and background attacks (Machanavajjhala *et al.*, 2007). However, there are downsides to ℓ-diversity. For starters, it may be difficult and not necessary, say if the sensitive attribute is binary like in the case of sex – male and female. The problem is exacerbated when there is a small number of one of the binary sensitive attributes (Li *et al.*, 2007). In addition, ℓ-diversity may not be enough to prevent attribute disclosure in cases where the distribution of the sensitive variables is skewed or the sensitive variables are very similar to each other e.g., income values.

4.2.4   *t*-Closeness

t-closeness provides privacy in cases where K-anonymity and ℓ-Diversity fail to do so (Li *et al.*, 2007). *t*-closeness requires the distribution of the sensitive values in each equivalence class to be "close" to the corresponding distribution in the original table.

In other words, as presented by Li *et al.* (2007), *t*-closeness is based on the premise that a user has prior knowledge of the sensitive attributes of a record. By using the prior knowledge about the individual record's sensitive attributes (β_0), along with prior knowledge of the distribution of those sensitive attributes in the population, the user can form a belief for that individual record (β_1). Then, once the user has access to the released table, they can use their knowledge to identify the corresponding class the record is in and to learn more about the distribution of those sensitive attributes in that class. This provides the user with more information on the individual record (β_2).

t-closeness aims to reduce the difference between β_1 and β_2. To do so, it is assumed that the distribution of the sensitive attributes in the population is public knowledge. With *t*-closeness, information is released in such a way that a user can learn very little additional information about an individual record. This means that the goal of *t*-closeness, is to have the distribution of sensitive information in the population and that of any class, as close as possible.

Li *et al.* (2007) state that the *t*-closeness principle states is:

> *"An equivalence class is said to have t-closeness if the distance between the distribution of a sensitive attribute in this class and the distribution of the attribute in the whole table is no more than a threshold t. A table is said to have t-closeness if all equivalence classes have t-closeness."*

Methods to measure the difference in distribution include variational distance, Kullback-Leibler distance, and Earth Mover's (EMD) distance.

t-closeness takes as input a table or data set $T(A_1,A_2\dots A_N)$, a parameter K specifying the minimum cluster or group size and a value for *t*. The output is a table , a set of clusters/groups satisfying K-anonymity and *t*-closeness. It works as follows (example from Soria-Comas *et al.* (2015) the Standard Microaggregation and Merging algorithm measures the distance using EMD):

1. Apply microaggregation to t with minimum cluster size K, store the output in T'.
2. While distance between T' and T is larger than t:
    a. Choose the cluster in T' with the greatest distance w.r.t. t and store this cluster in C.
    b. Choose the cluster in T' closest to C in terms of key variables, store this cluster in C'.
    c. Merge C and C' in T'.

Continuing with the same example but add more categories to the sensitive variable Disease.

Table 14   Table of medical condition data that has 0.25 -closeness with respect to Disease

	Non-sensitive			Sensitive
	Postal Code	Age	Occupation	Medical Condition
1	A1*	≤30	*	Heart Disease
2	A1*	≤30	*	Diabetes
3	A1*	≤30	*	Pneumonia
4	A1*	≤30	*	Bronchitis
5	C**	≥30	*	Cancer
6	C**	≥30	*	Diabetes
7	C**	≥30	*	Heart Disease
8	C**	≥30	*	Pneumonia
9	C**	≤50	*	Heart Disease
10	C**	≤50	*	Bronchitis
11	C**	≤50	*	Cancer
12	C**	≤50	*	Diabetes

Consider the sensitive variable of Table 14. The values of Disease are given as:

$$Q' = \{\textit{Heart Disease, Diabetes, Pneumonia, Bronchitis, Cancer, Diabetes, Heart Disease, Pneumonia, Heart Disease, Bronchitis, Cancer, Diabetes}\}.$$

In the set Q, there are three instances of *heart disease*, three instances of *Diabetes*, two instances of *Pneumonia*, two instances of *Bronchitis,* and finally two instances of *Cancer.* Accordingly, the distribution of these categories over Table 14 is $Q=\{\frac{3}{12}HD,\frac{3}{12}D,\frac{2}{12}P,\frac{2}{12}B,\frac{2}{12}C\}$. In the first equivalence class of Table 14: P_1 {*Heart Disease, Diabetes, Pneumonia, Bronchitis*} there is one occurrence of each of "*heart disease*", "*Diabetes*", "*Pneumonia*" and "*Bronchitis*" yielding a distribution of $P_1=\{\frac{1}{4},\frac{1}{4},\frac{1}{4},\frac{1}{4},0\}$. Given that the Disease is a categorical variable, we will apply the EMD distance. We recall that EMD stands for Earth Mover's Distance: $E\ (P,Q)=\frac{1}{2}\Sigma_{i=1}^{m}|p_i-q_i|$, a measure that is equal to one-half of the Manhattan distance.

So, t-closeness for the Disease is calculated as follows:

$$E\ (P_1,Q) = \tfrac{1}{2}[\ |\tfrac{1}{4} - \tfrac{3}{12}| + |\tfrac{1}{4} - \tfrac{3}{12}| + |\tfrac{1}{4} - \tfrac{2}{12}| + |\tfrac{1}{4} - \tfrac{2}{12}| + |0 - \tfrac{2}{12}|\ ] \approx 0.166.$$

Then, $P_2=\{\frac{1}{4},\frac{1}{4},\frac{1}{4},0,\frac{1}{4}\}$ and $E\ (P_2,Q) = \tfrac{1}{2}[\ |\tfrac{1}{4} - \tfrac{3}{12}| + |\tfrac{1}{4} - \tfrac{3}{12}| + |\tfrac{1}{4} - \tfrac{2}{12}| + |0 - \tfrac{2}{12}| + |\tfrac{1}{4} - \tfrac{2}{12}|\ ] \approx 0.166.$

Finally, $P_3=\{\frac{1}{4},\frac{1}{4},0,\frac{1}{4},\frac{1}{4}\}$ and $E\ (P_3,Q) = \tfrac{1}{2}[\ |\tfrac{1}{4} - \tfrac{3}{12}| + |\tfrac{1}{4} - \tfrac{3}{12}| + |0 - \tfrac{2}{12}| + |\tfrac{1}{4} - \tfrac{2}{12}| + |\tfrac{1}{4} - \tfrac{2}{12}|\ ] \approx 0.166.$

Thus, $t = \max(0.166, 0.166, 0.166) = \mathbf{0.166}.$ For more detail on t-closeness, see Dosselmann *et al.* (2019).

4.3　Other methods for disclosure evaluation

4.3.1　Peer Review

Peer review of disclosure methods, such as rounding, generalisation, suppression, or even differential private methods such as Laplace's mechanism, is a fit for purpose exercise to determine and demonstrate whether or not confidentiality methods or privacy-preserving techniques are fit for use based on an NSO's own legislative and operational frameworks.

For example, Statistics New Zealand's disclosure control practices are based on New Zealand statute, international and national best practices. Disclosure risk according to Statistics New Zealand is based on these objectives:

- Maintaining privacy via confidentiality, by outputting zero 'sufficiently accurate' disclosures of individuals, or of particulars relating to individuals.
- Maintaining data quality via confidentiality, by using methods which introduce zero bias, or as little bias as possible, into the original data, and the original data's means and other data set measures, where these means and measures are estimated using published 'confidentialised' data.

In practice, Statistics New Zealand endeavours to have zero tolerance for publishing accurate or discernible disclosures of counts of 1 or 2, or particulars relating to one- or two-unit records. This includes also potentially not disclosing counts of 0, and/or counts of 3, 4, or 5, as 'coverage' or 'protection' for counts of 1 or 2, etc.

Some typical methods to achieve this end, either used by or under investigation for use at Statistics New Zealand, and which demonstrably introduce limited bias, include:

1. Random rounding to base 3.
2. Fixed, or consistent, random rounding to base 3.
3. P% rule-based suppression and aggregation.
4. Noised Counts and Magnitudes (NCM), which is also substantially a differentially private method, and hence could also be measured via its differential privacy parameters.
5. R-*synthpop* Classification and Regression Tree (CART) based non-1:1-mapped synthetic data, preserving univariate and bivariate inferential validity, which is also substantially a differentially private method, and hence could also be measured via its differential privacy parameters.

In summary, in terms of measuring disclosure risk, confidentiality methods which do not accurately disclose counts of 1 or 2, or particulars relating to one- or two-unit records, and which do not introduce bias, are fit for purpose.

International best practices for peer review can be found in the European Statistical System Code of Practice Peer Reviews: The National Statistical Institute's guide, Version 1.3 (Eurostat, 2007).

4.3.2   Feature mean Scaled Variance

A measure of disclosure risk suitable for data with a one-to-one mapping between the original and synthetic data is call *feature mean scaled variance*. This method is only suitable and fit-for-purpose for data with a 1:1 mapping between original data and synthetic data; and which is all ordinal data, e.g., Boolean variables, ordinal numerical variables, ordinal categoric variables.

In this measure, all variances between mapped original and synthetic data points are feature scaled (or normalised) to the range 0 and 1. These features scaled variances substantively report combinable inaccuracy values for the synthetic data compared to the original data. Value of around 0 refers to identical data. Values of around 0.5 or more refer to highly non-identical data.

Let's look at an example of fictitious data in Table 15.

Table 15   Example of original data

Record index	Field 1	Field 2	Field 3	Field 4	Field 5	Field 6	Field 7	Field 8	Field 9	Field 10
1	1	2	5	10	100	0.5	0.2	0.1	-1	-0.5
2	2	4	10	20	200	1	0.4	0.2	-2	-1
3	3	6	15	30	300	1.5	0.6	0.3	-3	-1.5
4	4	8	20	40	400	2	0.8	0.4	-4	-2
5	5	10	25	50	500	2.5	1	0.5	-5	-2.5
6	6	12	30	60	600	3	1.2	0.6	-6	-3
7	7	14	35	70	700	3.5	1.4	0.7	-7	-3.5
8	8	16	40	80	800	4	1.6	0.8	-8	-4
9	9	18	45	90	900	4.5	1.8	0.9	-9	-4.5
10	10	20	50	100	1000	5	2	1	-10	-5

The first step in calculating the feature mean scale variance is to normalise record 1 (x_1) in field 1:

$$\frac{x_1 - \min{(field1)}}{\max{(field\ 1)} - \min{(field\ 1)}}$$

We do this for each record in each field and get the normalised values found in Table 16. We then conduct the same activity for the synthetic data, where Table 17 contains the synthetic data, and table 18 contains the normalised synthetic data. We then take the absolute difference of each record's normalised values between the synthetic and original data. This then would produce Table 19, which is the feature mean scale variance between Tables 16 and 18.

An arbitrary safety floor threshold for publication purposes for this method is proposed to be in the value range (0.05, 0.2). In other words, for each unit record row in Table 19, if the average of fields {1…10} for that row is greater than the arbitrary safety floor threshold you choose in the value range (0.05, 0.2), then the corresponding synthetic unit record row in Table 17 is considered by this method to be sufficiently safe to release.

Table 16 Normalised version of Table 15

Record index	Field 1	Field 2	Field 3	Field 4	Field 5	Field 6	Field 7	Field 8	Field 9	Field 10
1	0.1	0.1	0.1	0.1	0.1	0.1	0.1	0.1	0.9	0.9
2	0.2	0.2	0.2	0.2	0.2	0.2	0.2	0.2	0.8	0.8
3	0.3	0.3	0.3	0.3	0.3	0.3	0.3	0.3	0.7	0.7
4	0.4	0.4	0.4	0.4	0.4	0.4	0.4	0.4	0.6	0.6
5	0.5	0.5	0.5	0.5	0.5	0.5	0.5	0.5	0.5	0.5
6	0.6	0.6	0.6	0.6	0.6	0.6	0.6	0.6	0.4	0.4
7	0.7	0.7	0.7	0.7	0.7	0.7	0.7	0.7	0.3	0.3
8	0.8	0.8	0.8	0.8	0.8	0.8	0.8	0.8	0.2	0.2
9	0.9	0.9	0.9	0.9	0.9	0.9	0.9	0.9	0.1	0.1
10	1	1	1	1	1	1	1	1	0	0

Table 17 Synthetic version of Table 15

Record index	Field 1	Field 2	Field 3	Field 4	Field 5	Field 6	Field 7	Field 8	Field 9	Field 10
1	1	2	5	10	100	0.5	0.2	0.1	-1	-0.5
2	0	4	10	20	200	1	0.4	0.2	-2	-1
3	0	0	15	30	300	1.5	0.6	0.3	-3	-1.5
4	0	0	0	0	0	2	0.8	0.4	-4	-2
5	2.5	5	12.5	25	250	1.25	0.5	0.25	-2.5	-1.25
6	0.3	1.8	7.5	21	270	1.65	0.78	0.45	-5.1	-2.85
7	0	3.5	17.5	52.5	700	0	0.35	0.35	-5.25	-3.5
8	7.2	14.4	36	72	720	3.6	1.44	0.72	-7.2	-3.6
9	7.2	14.4	36	72	720	3.6	1.44	0.72	-7.2	-3.6
10	0	0	0	0	0	0	0	0	0	0

Table 18 Normalised version of Table 17

Record index	Field 1	Field 2	Field 3	Field 4	Field 5	Field 6	Field 7	Field 8	Field 9	Field 10
1	0.1	0.1	0.1	0.1	0.1	0.1	0.1	0.1	0.9	0.9
2	0	0.2	0.2	0.2	0.2	0.2	0.2	0.2	0.8	0.8
3	0	0	0.3	0.3	0.3	0.3	0.3	0.3	0.7	0.7

Table 18 Normalised version of Table 17 (continued)

Record index	Field 1	Field 2	Field 3	Field 4	Field 5	Field 6	Field 7	Field 8	Field 9	Field 10
4	0	0	0	0	0	0.4	0.4	0.4	0.6	0.6
5	0.25	0.25	0.25	0.25	0.25	0.25	0.25	0.25	0.75	0.75
6	0.03	0.09	0.15	0.21	0.27	0.33	0.39	0.45	0.49	0.43
7	0	0.175	0.35	0.525	0.7	0	0.175	0.35	0.475	0.3
8	0.72	0.72	0.72	0.72	0.72	0.72	0.72	0.72	0.28	0.28
9	0.72	0.72	0.72	0.72	0.72	0.72	0.72	0.72	0.28	0.28
10	0	0	0	0	0	0	0	0	1	1

Table 19 Absolute difference of normalised values or feature mean scale variance between synthetic and original data

Record index	Field 1	Field 2	Field 3	Field 4	Field 5	Field 6	Field 7	Field 8	Field 9	Field 10	Average Variance
1	0	0	0	0	0	0	0	0	0	0	0
2	0.2	0	0	0	0	0	0	0	0	0	0.02
3	0.3	0.3	0	0	0	0	0	0	0	0	0.06
4	0.4	0.4	0.4	0.4	0.4	0	0	0	0	0	0.2
5	0.25	0.25	0.25	0.25	0.25	0.25	0.25	0.25	0.25	0.25	0.25
6	0.57	0.51	0.45	0.39	0.33	0.27	0.21	0.15	0.09	0.03	0.3
7	0.7	0.525	0.35	0.175	0	0.7	0.525	0.35	0.175	0	0.35
8	0.08	0.08	0.08	0.08	0.08	0.08	0.08	0.08	0.08	0.08	0.08
9	0.18	0.18	0.18	0.18	0.18	0.18	0.18	0.18	0.18	0.18	0.18
10	1	1	1	1	1	1	1	1	1	1	1

One also has to choose another arbitrary utility ceiling threshold for the same unit records. This could be in the value range (0.2, 0.5); but work on an appropriate arbitrary utility ceiling threshold range for this method remains an open research question. But, in other words, for each unit record row in Table 19, if the average of fields {1…10} for that row is less than the arbitrary utility ceiling threshold you choose in the value range (0.2, 0.5), then the corresponding synthetic unit record row in Table 17 is considered by this method to be of sufficient utility to release.

Exact thresholds here are a matter of risk appetite, both from a safety perspective and from a utility perspective. By analogy, Abowd (2016) suggests approximately 90% accuracy and 10% inaccuracy could be an appropriate target, hence how an arbitrary safety floor threshold of 0.1 is within the proposed safety floor threshold range (0.05, 0.2).

4.4   Tips to get started

Disclosure risk for synthetic data is still an area of development in the synthetic data community. Disclosure measures, risk, and tolerances are very organizational specific, context dependent, and based on NSO legal and operational frameworks. This makes it difficult at this point in time to make definitive recommendations on disclosure risk thresholds and methods.

Based on your own organization's disclosure legislation, frameworks and practices, Figure 10 can guide the synthesiser on any disclosure measures and techniques they choose to apply to their synthetic data.

Figure 10   Decision tree to aid in disclosure control considerations

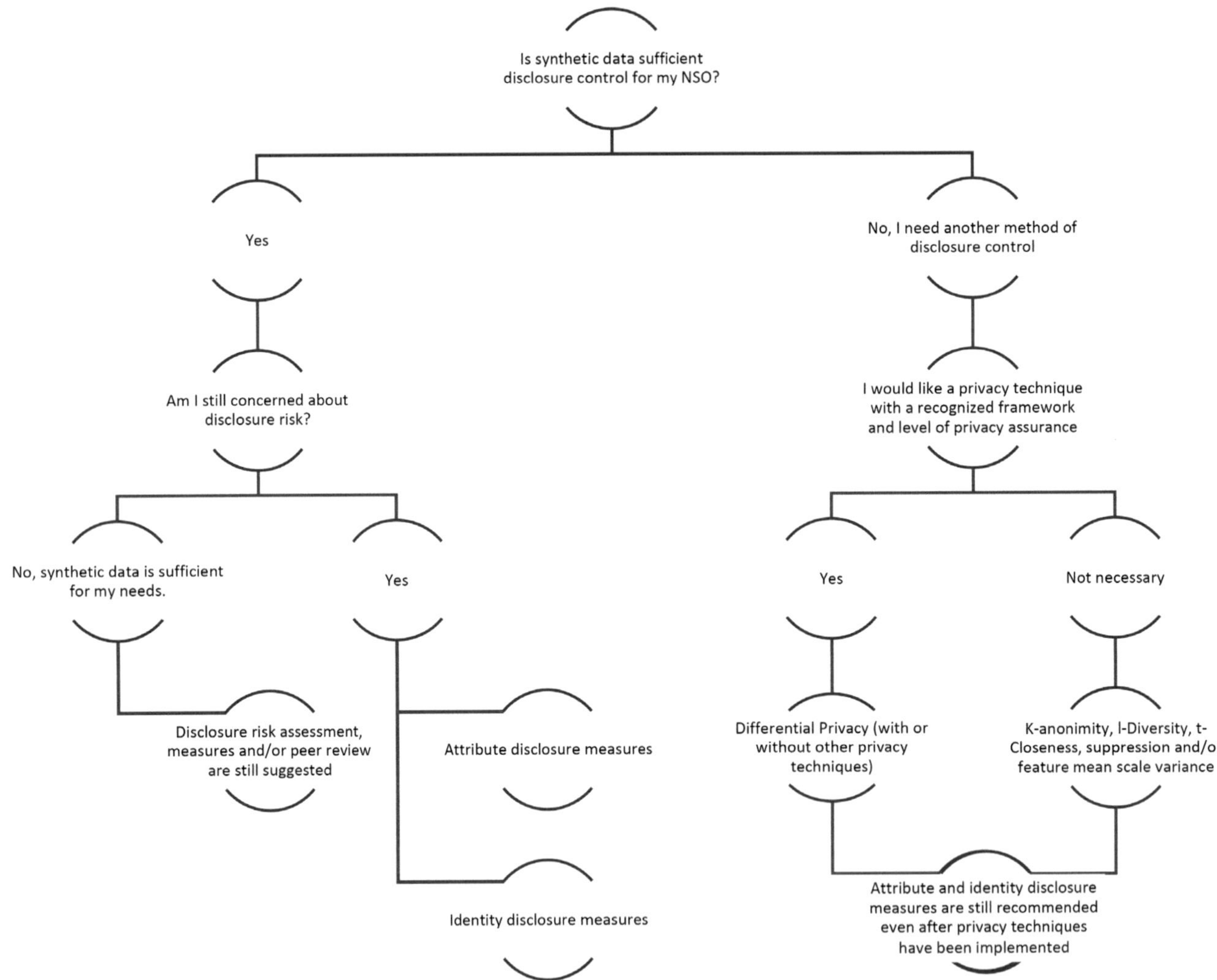

If you decide to implement additional privacy preserving techniques, attribute disclosure and identity disclosure measures are still useful to evaluate the effectiveness of the techniques. Any disclosure measures or techniques must also be implemented with the utility of the data in mind. Methods to measure and evaluate the utility of synthetic data, with or without additional privacy techniques are discussed in the next chapter.

A variety of disclosure measures and privacy preserving techniques were explored in the *HLG-MOS Synthetic Data Challenge 2022* (Bhagat *et al.*, 2022).[18] Though few of these methods have been formally implemented in NSOs, the measures and techniques from the challenge can serve as a baseline for future research on disclosure measures for synthetic data.

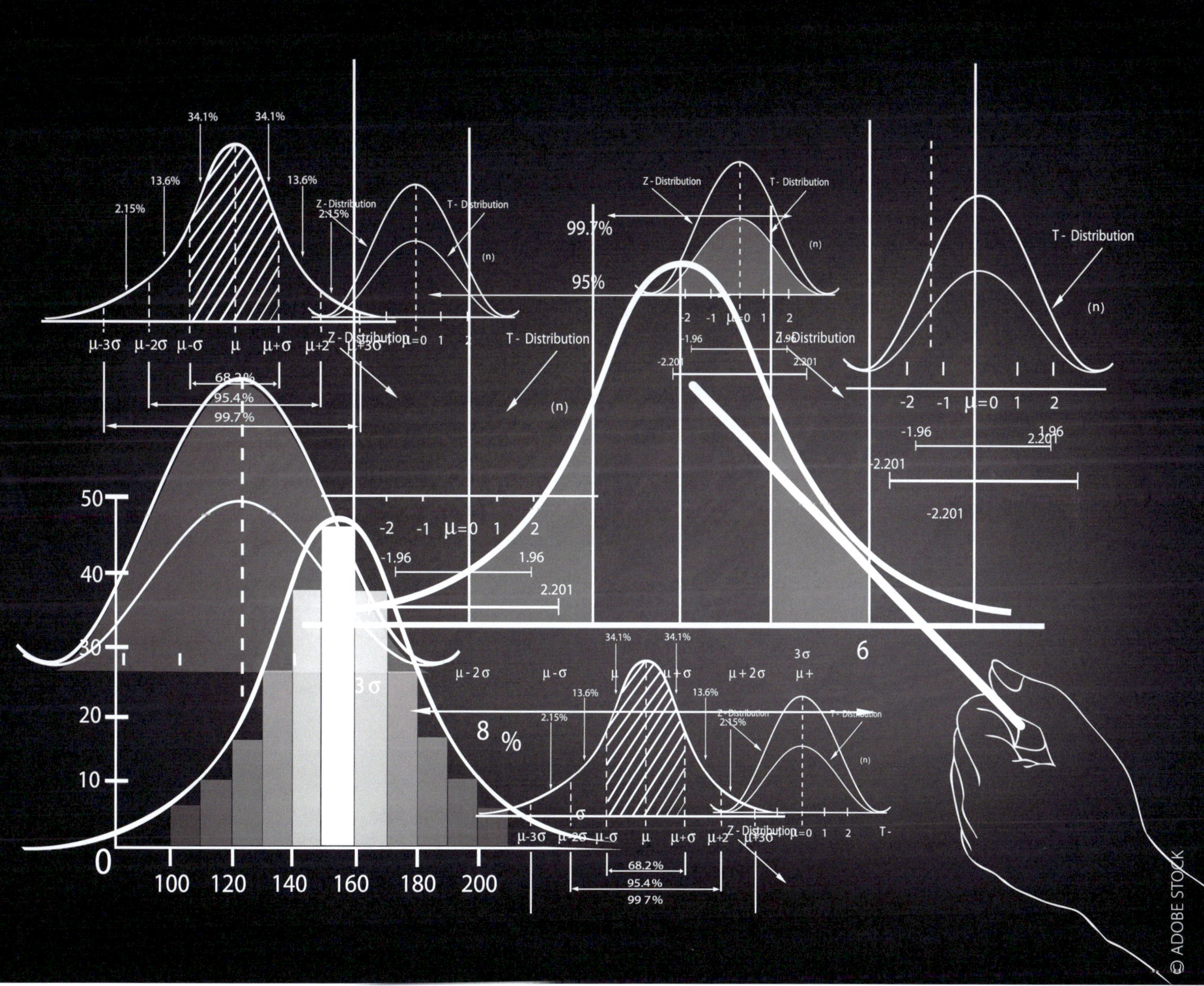

---

18   See https://pages.nist.gov/HLG-MOS_Synthetic_Data_Test_Drive/index.html#privacy_evaluation_methods for
      a summary and results of the privacy evaluation measures and techniques used in the HLG-MOS Synthetic Data
      Challenge 2022.

Chapter 5: Utility measures for evaluating synthetic data

The utility, or value, of a synthetic data set reflects how useful that data set is to the purpose or the use case for the data. As discussed in Chapter 2, synthetic data is often used either instead of the original data, or as a preliminary analysis to guide the final results which will be run on the original data. In both cases, the utility of synthetic data is based on how similar the conclusions are between the synthetic data and the original (confidential) data. This is equally important in the latter use case because preliminary analysis on the synthetic data will guide the final models used.

This guide recommends in Chapter 3 that the methods of creating synthetic data should depend on the use case, what the synthesiser wants to preserve and the type of original data. Once the synthetic data sets have been created, the utility can be evaluated.

There are two broad categories of utility measures: "broad", "global" or "general" utility measures, as opposed to "narrow" or "specific" measures. In this guide, we will use the terms *general* and *specific* measures. Specific measures are useful when evaluating a specified analysis, however they are not useful for tuning – modifying synthesis methods to improve the utility – as the synthesiser often does not know the analysis that will be conducted using the synthetic data when tuning has to take place.

According to Raab *et al.* (2021), there are two main reasons we might wish to evaluate the general utility of synthetic data:

1. To compare different synthesis methods for the same data set in order to generate the most useful synthetic data set for the user.
2. To diagnose where the original and synthetic data distributions differ and thus tune the synthesis methods to improve the utility of the synthetic data.

For the first of these, a number of measures have been proposed that summarise the utility of the data, or sometimes of a subset of the data, by a single number. Two main methods have been proposed to compute these measures. The first, proposed by Karr *et al.* (2006) and Woo *et al.* (2009), is to combine the two data sets (original and synthetic) and to use the information in the records to predict their source. Several measures can be calculated from this approach some of which are calculated from the propensity score, the probability that a record is from the synthetic data. The second method used to compute a single comparative measure is to compare tables created from the synthetic data with those from the original. These two methods are related, as we will discuss below.

A single measure does not provide guidance as to what aspects of the synthetic data differ from the original. Thus, we need different strategies to fulfil the second of Raab's requirements. Several methods have been suggested for this, sometimes making use of the utility measures discussed above for subsets of variables in the records.

Table 20   Summary of utility measures

Method	Measure or procedure	Reference	Measure acronym
Specific	Confidence interval overlap	Karr et al. (2006)	
	Achieving specific outcomes	Kaloskampis et al. (2020), Jordon et al. (2018) and Slokom et al. (2021)	
General Single measure from propensity score	Propensity *score* mean squared error	Karr et al. (2006) and Woo et al. (2009)	pMSE
	Kolmogorov- Smirnov Statistic comparing propensity scores for original and synthetic data	Bowen et al. (2021)	SPECKS
	Other comparisons of propensity scores for original and synthetic data e.g., Wilcoxon signed rank statistic	Raab et al. (2021)	U
	Percentage over 50% of combined records correctly predicted by the propensity score	Raab et al. (2021)	PO50
General Single measure from tables	Voas-Williamson statistic	Voas and Williamson (2001)	VW
	Freeman-Tukey (FT)	Voas and Williamson (2001)	FR
	Likelihood ratio statistic from tables and other members of the divergence family	Voas and Williamson (2001)	G
	Jensen-Shannon Divergence	Fuglede and Topsoe (2004)	JSD
	Bhattacharyya metric	Bhattacharyya (1943)	dBhatt
	Mean absolute difference in densities	Raab et al. (2021)	MabsDD
	Difference of correlation matrices	Kaloskampis et al. 2019	
	Weighted mean absolute difference in densities	Raab et al. (2021)	WMabsDD
Methods for exploring utility	Comparing histograms by visualisation and summary measures	http://www.synthpop.org.uk and Kaloskampis et al. (2020)	
	Comparing cross-tabulations from marginal distributions	Raab (2011) and NIST (2021)	
	Comparing and visualising other summary statistics (e.g. Pearson Correlations)	Beaulieu-Jones et al. (2019) and Kaloskampis et al. (2020)	

Table 20 provides a summary of the methods discussed in this chapter classified as methods for specific utility, as single measures of general utility, either by propensity score or by tables, or as methods for exploring, summarising or visualising utility.

5.1   Specific utility measures

Specific utility measures compare the results of statistical models fitted to the synthetic and the original data. To begin with, any statistical analysis can be used to create a utility measure, for example, difference in means of variables, differences in correlation, tables and cross-tabulations and even resulting analysis or outcomes. The most widely used of these is the confidence interval overlap that provides both a summary measure and a visualisation of the results from a statistical model from the two data sources. Other summary measures include various graphical comparisons as well as standardised differences in coefficients and an overall lack of fit measure that can be computed from the variance matrix of the coefficients.

5.1.1   Confidence interval overlap

As discussed in Chapter 2, one of the most popular use cases for synthetic data is testing analysis, where often a user of synthetic data is testing a linear or generalised linear regression. This activity not only produces coefficients but also confidence intervals. A measure to assess the utility of synthetic data for testing analysis is to evaluate how the confidence intervals of an estimate differ between the original and synthetic data (Karr *et al.*, 2006). The confidence interval overlap measure is such a measure. Karr *et al.* (2006) suggest using the percentage overlap of confidence intervals (IO), defined for each coefficient β_i as,

$$IO_i = 0.5 \left[\frac{\min(u_o,u_s)-\max(l_o,l_s)}{u_o-l_o} + \frac{\min(u_o,u_s)-\max(l_o,l_s)}{u_s-l_s} \right]$$

where the confidence interval for the original data is $(u_o; l_o)$ and for the synthesised data $(u_s; l_s)$. The numerators in each of the terms in this equation are the overlap of the intervals which becomes negative when the intervals are disjoint. The average of the overlaps can then be used as a summary measure of utility. The IO measure takes a maximum value of 1 when the intervals are the same length, but lower when they are different lengths.

Figure 11   Illustration of confidence interval overlap from logistic regression.

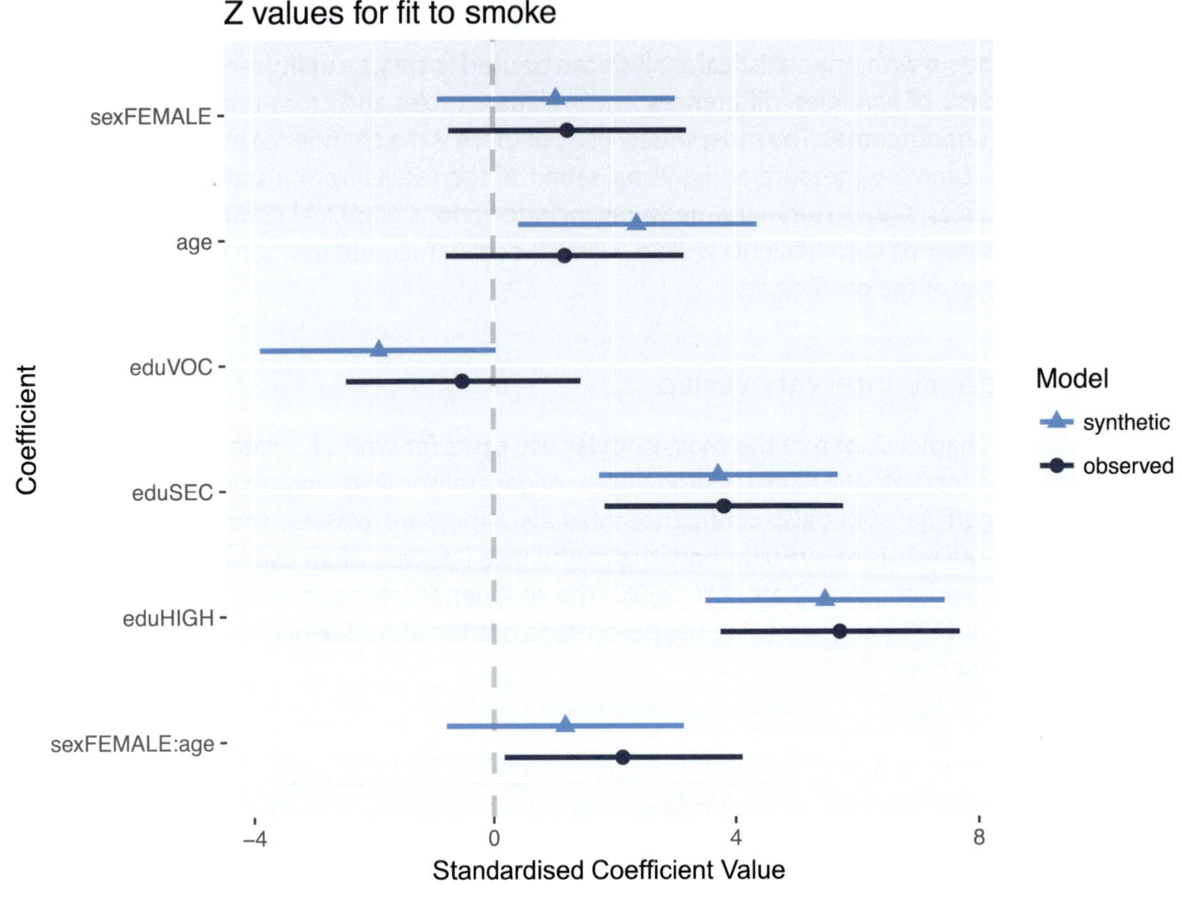

Source: Raab and Nowok (2017)

Confidence interval overlaps are most often used to compare results for fitting statistical models to the original and synthetic data. It is recommended that a graphical display that compares the fit of the models should be the first step in evaluating the fits from the two sources. An example is Figure 11: Illustration of confidence interval overlap from logistic regression, from Raab and Nowok (2017) which shows the output from the *synthpop* package for R (www.synthpop.org.uk).

5.1.2   Achieving specific outcomes

A key characteristic that synthetic data should have in order to be useful is for the results of the specific analysis or task at hand to be the same with the synthetic data as it is with the original data. These methods are only useful when that specific analysis or task is known. In its simplest form, a synthesiser would want to identify tasks relevant to the data set (e.g., a classification task), and then compare the task accuracy between the original and synthetic data. Task accuracy is a common method to evaluate deep learning models, where the same classification or decision needs to be implemented for the original and synthetic data. This measure is considered specific because it evaluates utility for a specific rather than a generic task. Task accuracy measures are useful because there can be cases for which more general utility measures indicate favourable levels of utility, but where results from the synthetic data may still not be the same as the original data for a particular task.

For example, take the Adult Income data set from UCI repository that has 32,500 records of Americans with variables that include age, working status, education, and income (Kohavi and Becker, 1996). With this data, the task at hand is to see if income is above or below a certain amount, say make a prediction if an income is above or below $50,000. The first step is to build a model on the original data and determine the accuracy of the model. Next, conduct the same task with synthetic data. The same exercise on the original and synthetic data may result in different outcomes, which provides the synthesiser an assessment of synthetic data accuracy. Figure 12 illustrates such a classification task with the original data (red) and with synthetic data generated using GANs and three values of privacy loss (blue) (Kaloskampis *et al.*, 2020). As shown in Figure 12, the synthesiser or user can now assess if the task using the synthetic has the utility necessary for their use case.

Figure 12 Classification accuracy trained on original Adult Income data set and synthetic data sets generated with GANs, with different values of privacy loss ε

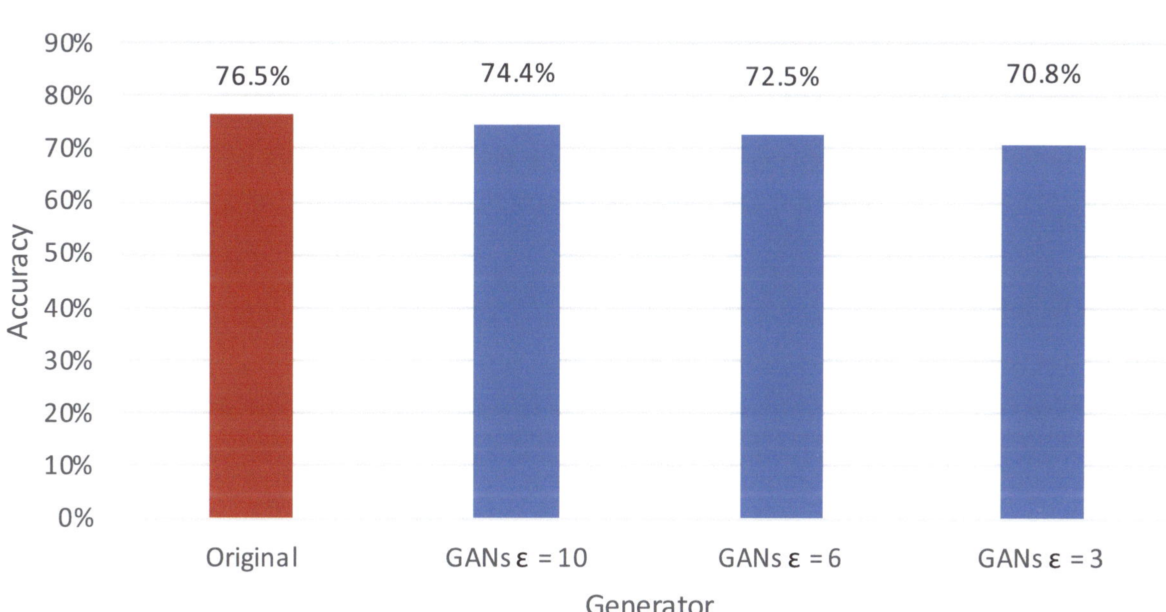

Source: Kaloskampis et al. (2020).

Alternatively, a synthesiser may be interested in a broader evaluation and look at the performance of a machine learning algorithm as a whole. Such an evaluation involves comparing the performance metrics of predictive models trained on synthetic and on original data (called model compatibility).

This performance of machine learning models trained and tested on original and/or synthetic data is compared based on different scenarios (Heyburn *et al.*, 2018, Jordon *et al.*, 2018, Fekri *et al.*, 2020, Slokom *et al.*, 2021):

- Train on Real and Test on Synthetic data ($TRTS$),
- Train on Synthetic and Test on Real ($TSTR$),
- Train on Real, Test on Real ($TRTR$),
- Train on Synthetic, Test on Synthetic ($TSTS$),
- Lastly, trained and tested on a mixture of real and synthetic data ($TMTM$).

In principle, these scenarios are transferable to the evaluation of synthetic data. However, it is important to consider whether $(\mathcal{TRTS})$ and $(\mathcal{TSTR})$ actually yield meaningful information about how useful synthetic data is for a specific purpose. The reason is that, if the synthetic data provides synthetic users, then users in the training set (or test set) are different from those in the test set (respectively training set). So, it is critical to develop evaluation frameworks that are suitable for use in evaluating synthetic data.

Inspired by notations proposed by Jordon *et al.* (2018), let's consider a data set, D, a task t, that is going to be performed on D. We split D into training set, *DTrain*, and testing set, *DTest*. Let A1..Ai..,AN be N machine learning algorithms that take as input a training set *DTrain*, and provides as output a prediction model *Ai (DTrain)*. Also, we consider mt a performance metric for task t that takes as input a trained model, $\mathcal{M}$, and a testing data set, *DTest*, and outputs a value in $\mathbb{R}$. $\mathcal{G}$ denotes a synthetic data generation method and $D\mathcal{G}$, denotes the synthetic data set generated by $\mathcal{G}$. As for the original data, we will split the synthetic data into a training set, $D\mathcal{G}$Train , and testing set, $D\mathcal{G}$Test.

Jordon *et al.* (2018) proposed a metric called synthetic ranking agreement (for short SRA). SRA can be formulated as follows:

$$mt(\text{Ai } (D\mathcal{G}\text{Train}), D\mathcal{G}\text{Test}) < mt(\text{Aj } (D\mathcal{G}\text{Train}), D\mathcal{G}\text{Test})$$
$$=> mt(\text{Ai } (D\text{Train}), D\text{Test}) < mt(\text{Aj } (D\text{Train}), D\text{Test})$$

For all i, j $\in$ {1, ..., N}, i$\neq$j.

Thus, synthetic ranking agreement (SRA) is defined as:

$$\mathbf{SRA}(\mathcal{G}) = \frac{1}{k(k-1)} \sum_{i=1}^{k} \sum_{j \neq i} \mathbb{I}\Big((R_i - R_j) \times (S_i - S_j) > 0\Big)$$

Where Ri $= mt(\text{Ai } (D\text{Train}), D\text{Test})$, and Si $= mt(\text{Ai } (D\mathcal{G}\text{Train}), D\mathcal{G}\text{Test})$, for each i = 1, .., N. So that Ri represents the performance of algorithm i on the original data and Si represents the performance of algorithm i on the synthetic data.

The SRA can be thought of as the (empirical) probability of a comparison on the synthetic data being "correct" i.e. the same as the comparison would be on the original data.

5.2   General utility measures giving a single measure

The statistical analyses for which the synthetic data will be used are typically not known when the synthesis is being carried out. Measures that compare the whole distribution of the synthetic data to that of the original data are referred to as general utility measures. As we mentioned in the introduction to this chapter, these are mostly based on two methods, firstly combining the original and synthetic data and calculating a propensity score, the probability that any record is synthetic, and secondly by comparing tables of original and synthetic data. As we will show below these two methods can be considered to be the same thing, but comparing tables allows some extra measures that are not available from the propensity score method.

5.2.1   Measures from the propensity score

Karr *et al.* (2006) and Woo *et al.* (2009) evaluate the utility of synthetic data by combining the records of the original and synthetic data and measuring how well the data values predict the source of the records as original or synthetic. An indicator, say x, is assigned a value of 1 for the synthesised data and 0 for the original. A method such as logistic regression or any non-parametric predictive method can be used to attempt to derive the propensity score, $\hat{p}$. Propensity score mean squared error measures the overall distribution similarity between the synthetic and original data. More specifically, the propensity score is the probability that x = 1, meaning that the record was from the synthesised data. If the distributions of the original and synthetic data are indistinguishable then all propensity scores are expected to be close to the proportion of synthetic records in the combined set; 0.5 if the two data sets have the same number of records. Several measures can be computed from $\hat{p}$, four of which are listed in Table 20. The most commonly used is the propensity score mean-squared error (pMSE) (Woo *et al.* (2009)).

To start calculating the pMSE, merge the original and synthetic data sets, adding an additional variable equal to one for all rows from the synthetic data set and equal to zero for all rows from the original data set. Second, for each record in the original and synthetic data, compute the probability of being in the synthetic data set, i.e. the propensity score. Any method to predict a binary variable is suitable for this method. Examples of such methods are: logistic regression, classification and regression trees, classification methods such a neural networks or random forest, or even from tables by calculating the proportions of synthetic to all records in each corresponding cell. Lastly, compare the distributions of the propensity scores in both data sets using

$$pMSE = \sum_{i} (\hat{p}_i - c)^2 / N$$

where c is the proportion of synthesised rows in the combined data = n_2/N (Raab *et al.*, 2021).

Table 20 presents many other propensity score based measures available in the literature and many synthetic data generation and evaluation packages such as synthpop. However, there are a few of these methods that are more frequently used in practice, namely propensity score mean squared error (S_pSME) as well as SPECKS and PO50.

SPECKS or Kolmogorov Smirnov statistic is the maximum distance between the cumulative distributions functions of the propensity score for the synthetic and original distributions. The formula for SPECKS based on propensity score measure is defined as:

$$SPECKS = sup_{\hat{p}} \left| F_{t=0}(\hat{p}_i) - F_{t=1}(\hat{p}_i) \right|$$

PO50 is the percentage above 50% of synthetic data records where the model used correctly predicts whether the record is original or synthetic data. The formula for PO50 is:

$$PO50 = 100 \sum_{i} [t_i\,(\hat{p}_i > c) + (1 - t_i)\,(\hat{p}_i < c)] / \sum_{i} (\hat{p}_i \neq c) - 50$$

where t_i is an indicator variable, i=1,…,N, taking the value 1 for rows from synthetic data and 0 for rows from original data.

5.2.2 Measures from tables

General utility measures can also be obtained from tables, as proposed by Voas and Williamson (2001) for summarising differences between synthetic data and the original. They adapted measures used in computing chi-squared tests for tables. In particular they suggested what we refer to as the Voas-Williamson statistic (VW) for comparing tables. It is similar to the usual Pearson chi-squared statistic (X^2). We can write the counts for any table with k categories as y_i (i = 1; 2; ..k) and the corresponding synthetic counts as s_i (i = 1; 2; ..k). If the total counts in each table are the same: $\Sigma y_i = \Sigma s_i = n$ then $X^2 = \Sigma_{i=1}^{k}(s_i - y_i)^2 / y_i$. A practical problem with this chi-squared statistic is that the contribution from a cell where the original data has a zero count, but is not a structural zero, is not defined.

Voas and Williamson propose the modification of this statistic by replacing y_i with the mean of y_i and s_i. Other statistics in the power-divergence family (Read and Cressie, 1988) could also be used, such as the deviance or the Freeman-Tukey measure (FT). Another related measure is the Jensen-Shannon Divergence (JSD) which can be considered as a modification of the likelihood ratio statistic to allow for zero values of y_i. Further measures computed from tables include the mean absolute difference in density (MabsDD) which is calculated as:

$$MabsDD = \frac{\Sigma_{i=1}^{k} \mid \frac{s_i}{n} - \frac{y_i}{n} \mid}{k}$$

Other measures include the histogram overlap measure, Bhattacharyya metric or dBhatt (Bhattacharyya, 1943).

All of these measures can be generalised to the case where the synthetic data has a different number of records from the original, see Raab *et al.* (2021) for details.

5.2.3 Relationships between the measures

Comparing tables can also be framed as a prediction measure where the propensity score in the case of equal sample sizes is just $\hat{p}_i = s_i /(s_i + y_i)$. Calculating this propensity score for an n-way table is identical to what would be obtained by using a logistic model with all (n-1)-way interactions. Several of the measures in Table 20 are linearly related. In particular, *VW* and *pMSE* are the same measure, as are the three measures *SPECKS, PO50* and *MabsDD* and also $dBatt \propto \sqrt{FT}$. Thus, there are fewer independent utility measures than Table 20 would suggest. Empirical investigations suggest that all of the measures are correlated with one another when compared for different synthetic data sets. For some subgroups, e.g., *VW/pMSE, FT, JSD* the correlations are so high as to suggest that they are essentially the same measure.

A small investigation of the ability of these measures to differentiate a poor synthesis from a good one suggested that *VW/pMSE, FT, JSD* performed slightly better than the other measures discussed here, but almost all gave satisfactory discrimination. These findings are based on recent empirical work by Raab et. al (2021), that would benefit from further development by other groups.

5.2.4   Scaling of utility measures

It is helpful if the utility measures can be on a scale that makes them easy to interpret. For all the measures described here a large value indicates a lack-of utility. One method could involve scaling the measures by the maximum value they could take. For example, JSD is scaled in this way since its maximum value is 1.0. Other measures have an interpretation that helps to understand them, for example, it is easy to think of the percentage correctly predicted for PO50, and dBhatt has an immediate interpretation as the overlap of matching histograms.

Another approach to scaling utility measures is to express them relative to the value that would be expected if the model used to synthesise the data was the "correct" model. The expected value for the "correct" model can be termed the Null expectation. This approach can also be considered as scaling the measures compared to the expected stochastic error of the distribution. The target value for measures scaled in this way is 1.0. All the measures derived from the various chi-squared tests have known Null expectations, Snoke *et al.* (2018) derived the Null distribution of this quantity for the pMSE when it is estimated from a model with a fixed number of parameters. They also propose methods for obtaining the Null expectation for any of these measures by replication methods. A modification of one of the methods to use for DP synthetic data has been proposed by Bowen *et al.* (2021). This scaling by the Null expectation differs from the others in that it defines a target that a good synthesis should achieve. Although an absolute target would be 1.0, synthetic data that have proved to be useful can have values in the range from 3 to 10. Values above 10 signal potential problems with some part of the distribution being evaluated. This is reasonable as we do not believe that original data are generated exactly from a statistical model.

5.2.5   Models for the propensity score

The choice of model for the propensity score is crucial to its performance and a more important choice than that of the utility measure. Any method that can predict group ownership could be used. Those that have been used in practice are logistic models, which include the special case of models that define tables, and classification and regression tree (CART) models. The models that can be fitted are limited by the complexity that it is possible to fit from a finite sample of data. Logistic models with a large number of parameters may fail to converge and, even if convergence is achieved, will have many parameters that cannot be estimated from lack of information (aliased parameters). Similarly, a comparison of tables with more than a small number of variables will yield large tables, with most of their cells having zero counts, that may lead to computational problems. CART models, that select a partition of the data to describe the distribution, can cope with data sets with more variables. But such models also have computational limits, especially when dealing with categorical variables with many possible levels, as are often found in data from NSOs. A simple model may give an assurance of a good fit although only a very limited aspect of the distribution differences has been assessed.

5.2.6   One number is not enough to describe utility

A person creating synthetic data needs more than a single number to assess the utility of the data they have produced. If the utility appears unsatisfactory, they need to know which aspects of the distribution are causing the problem. Many strategies can be devised to explore these differences. Some of these are described in the next section, some using the utility measures described here for subsets of variables or for the partitioning of large synthetic data sets into smaller strata, often defined by geographic areas.

5.3 Methods to explore aspects of utility

5.3.1 Univariate comparisons

The starting point of any evaluation of synthetic data is to examine how well the synthetic data reproduce the univariate distribution of each variable. Bar charts of categorical variables or histograms of numerical variables are the obvious first step. These may be accompanied by utility measures computed from the tabulation of each variable. The function to produce plots of each variable in the *synthpop* package can be accompanied by a table of a variety of utility measures. Kaloskampis *et al.* (2020) produce similar plots where they display the Bhattacharyya metric alongside the histograms, see Figure 13.

Figure 13 Comparison of histograms of the workclass variable of the Adult Income data set between original (red) and synthetic data sets (blue) generated with GANs, using different values of privacy loss ε. We denote the Bhattacharyya metric by dBhat

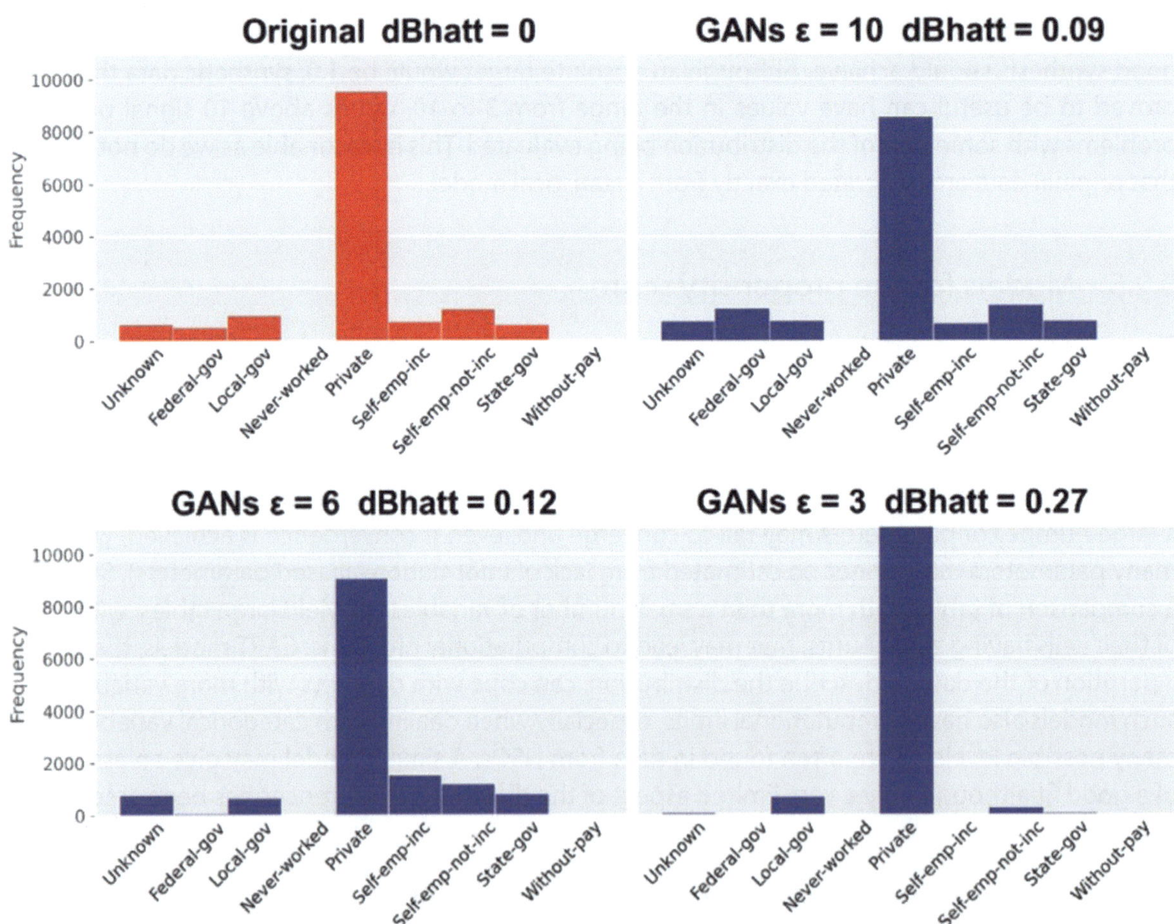

Source: Kaloskampis et al. (2020).

5.3.2   Marginal comparisons

Several methods have been proposed for comparing low-level marginals between the original and synthetic data. Any of the utility metrics that can be computed from tables can be used after first forming categories from any continuous variables. To summarise results from marginals, Raab *et al.* (2020) propose the following steps.

1. Examine histograms that compare the original and synthetic data and also check the pMSE-ratio for the univariate comparison of each variable.

2. Once these are satisfactory, continue by visualising the utility of all two-way relationships between variables.

3. If there is one variable of particular interest, for example, an outcome variable in an epidemiological study, then it might be worth checking and visualising all three-way relationships that involve that variable.

Figure 14 illustrates the output from step 2 for four syntheses of the same original data. Plot (a) is from a default parametric synthesis, showing that there was a problem with the variable "weight". Plots (b) and (c) show how reordering and stratifying the synthesis improves the utility. Plot (d) shows that synthesising from a CART model, with no adjustment, gives better utility than any parametric model.

Figure 14   Visualisations of the utility of all two-way relationships between variables.

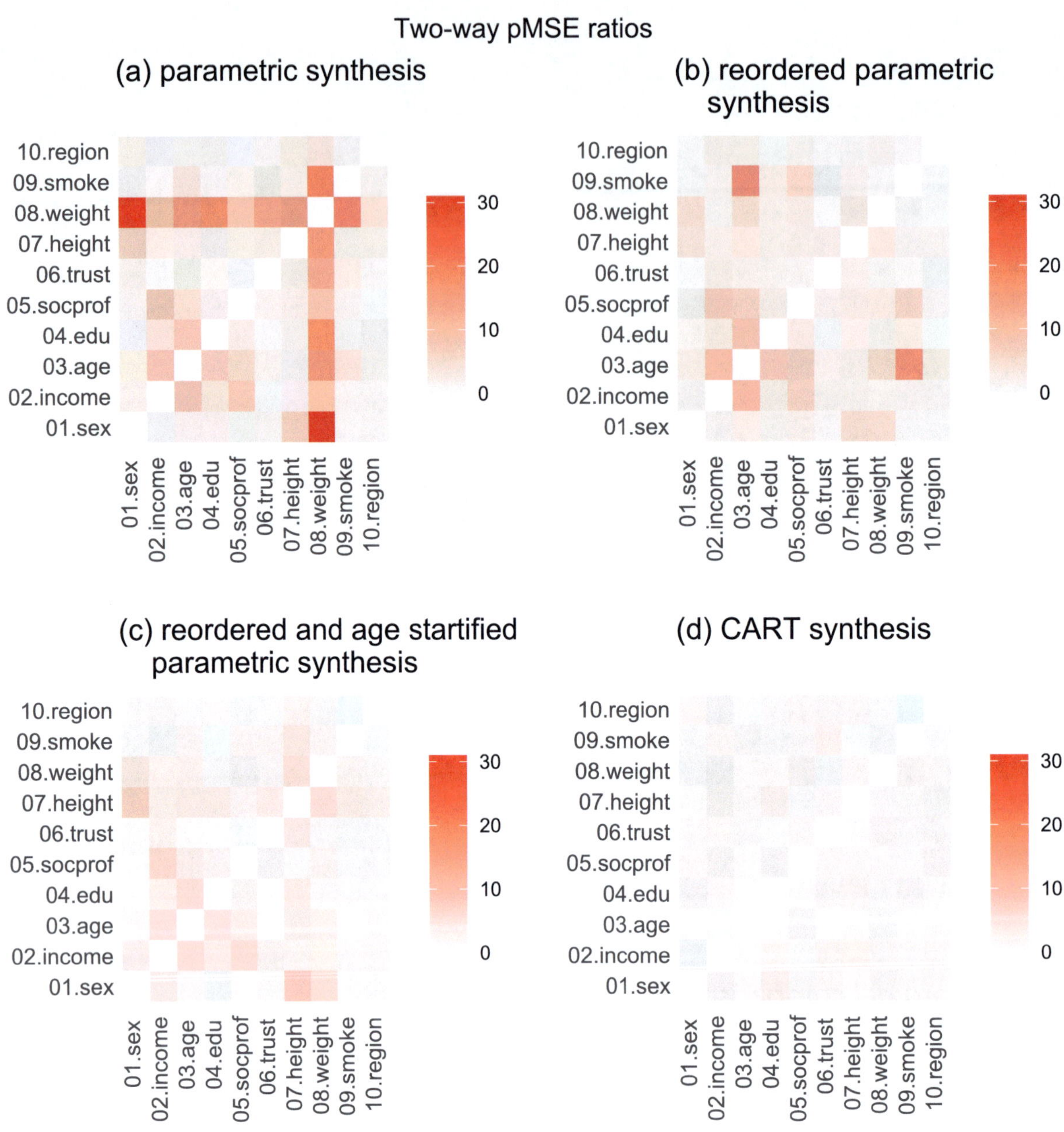

Source: Raab et al. (2021)

An example of marginal comparisons can be found in results from the 2018 National Institute of Standards and Technology's *Differential Privacy Synthetic Data Challenge*. During the challenge, a method based on three-way marginals to evaluate utility was evaluated. Marginal distribution metrics work well for discretised data, since one can easily consider all possible k-way margins of the full cross-classification of discrete variables. Numerical (integer or floating point) data can function under these metrics as well by discretisation. One implementation of a k-way marginal metric is to consider the total absolute deviation across cells of a marginal table based on two "versions" of a data set. After normalising the total absolute deviation by dividing the total of the table, one obtains a metric on how close the chosen margin is between the two data versions. This method is quite flexible and allows variations such as fixing certain dimensions to more finely assess differences for certain variables or constructs (Ridgeway *et al.*, 2021).

Diving into an example, we consider creating synthetic data for a subset of the 1940 Census Demonstration Data (Ruggles *et al.*, 2018) and assessing the synthetic data using k-way marginals. We take the subset of records for the District of Columbia, and synthesise the following variables:

- Binary sex (SEX)
- Age (AGE)[19]
- Major race category (RACE)
- Ethnicity (HISPAN)

We use two models for synthesis: a naïve model that simply permutes the values of each of the columns above (PERM), and a model based upon sequentially fit classification trees (TREE). The effect of both models is to produce a new data set that has the same scheme as the original but with potentially modified entries in each row. We then use marginal metrics to assess the relative quality of PERM and TREE in reproducing the original data.

A natural starting point, and indeed often the ending point for much exploratory data analysis, is to consider the one-way marginal metrics for each column. The margins for sex under each data model are presented in Table 21.

Table 21   The margins for sex under the PERM and TREE data models

	Original 1940	PERM	TREE
Male	318,269	318,269	317,917
Female	346,595	346,595	346,947

Source: Derived from Ruggles et al. (2018).

To get the sex marginal score under TREE, we take the absolute differences in the number of males and females under TREE, sum them, and divide by the total number of persons:

$$\frac{|317917 - 318269| + |346947 - 346595|}{318269 + 346595} = 0.00106$$

To get the overall 1-marginal score under TREE, we perform the same computation for the other three 1-way margins and then take the average with results shown in Table 22.

---

19   Though we synthesize the full range of ages, the marginal evaluations are on a recode of age into three bins
      for children (0-17), adults (18-64), and older adults (65+)

Table 22   Overall 1-marginal score under TREE model

Item	Value
Sex	1.06E-03
Age	8.00E-04
Race	3.55E-04
Ethnicity	1.99E-04
Average	6.03E-04

Source: Derived from Ruggles et al., 2018.

Since scores are more sensible when larger is better, we can use the transform:[20]

$$((2 - \text{raw\_score})/2) \times 1000$$

To map the raw scores to (1000, 0), so that now a score of 1000 means perfect recreation of the margin, and a score of 0 means the margin is maximally perturbed.[21]

We can then compare these overall 1-way marginal scores across our two models, as shown in Table 23.

Table 23   Comparison of 1-way marginal scores between models

Model	Raw Score	Adjusted Score
TREE	6.03E-04	999.7
PERM	0.00E+00	1000

Source: Derived from Ruggles et al., 2018.

The PERM method exactly recreates the 1-way margins since it samples the columns without replacement. But we see that TREE comes close, which becomes important when we consider marginals beyond one dimension.

We now consider the 3-way marginal metric. This is calculated exactly as above, except now we consider the deviations of the cells for all 3-dimensional margins. Here we see a change, as shown in Table 24.

---

[20]   Derived from the worst case: changing a table with counts of form [0 N] to [N 0] for a raw score of 2/N.

[21]   This score was used for the NIST 2018 Differential Privacy Synthetic Data Challenge: https://www.nist.gov/ctl/pscr/open-innovation-prize-challenges/past-prize-challenges/2018-differential-privacy-synthetic

Table 24   Comparison of 3-way marginal scores between models

Model	Raw Score	Adjusted Score
TREE	2.29E-03	998.85
PERM	3.61E-02	981.93

Source: Derived from Ruggles et al., 2018.

The TREE model now outperforms PERM. In this simple case, where the variables may not have especially strong dependencies, a naïve model such as PERM can still perform well, but as dependencies and dimensionality increase, this will not occur, and the job of the data synthesiser becomes a more delicate task.

We can extend marginal metrics to give finer detail. For instance, we can restrict margins to only those containing a set of variables (e.g., all tables that contain "AGE" as a margin). We can also consider the cell differences between different categories of a variable (e.g., compare scores restricted to cells associated with males versus those associated with females). In this way, we can build up a picture of where two data sets differ the most. This is especially helpful in assessing synthetic data when certain use cases may need preservation.

In the 2018 NIST *Differential Privacy Synthetic Data Challenge* example, all three-way marginals from their large data set would have been too many to compute, so a subsample of all possible marginals was used. Each was evaluated by calculating the MabsDD, that was then rescaled to give a "human-readable NIST score" defined as $1000\,(1 - MabsDD/2)$ that ranges from 0 to 1000, with 1000 representing exact agreement between the tables. The utilities can be examined to identify which variables contribute most often to the tables with low scores. This scoring method was also used to assess the utility of geographic subsets of the data.

5.3.3   Comparing other statistics

Many other statistics could be compared between the synthetic and original data. In their evaluation of the 2018 NIST synthetic data challenges, Bowen and Snoke (2021) propose a range of such measures as well as some of the other utility measures discussed here. They also propose methods of combining and visualising these different measures.

Beaulieu-Jones *et al.* (2019) have used a range of utility measures to evaluate DP synthetic data created from a clinical trial data set. They compare a number of relevant outcomes and present results graphically. In particular, they present the Pearson correlations (often referred to as correlation coefficient) between variables as heatmaps. Kaloskampis *et al.* (2020) have used the same method to present the correlations, as shown in Figure 15. As the visual comparison is often impractical, Kaloskampis *et al.* (2019) proposed a quantitative measure stemming from these visualisations, based on the difference of the underlying correlation matrices. This method could be useful, for example, in the process of hyperparameter optimisation of a synthetic data generation algorithm.

Figure 15 Pairwise Pearson correlation heatmaps for original Adult Income data set and synthetic data sets generated with GANs, with different values of privacy loss ε

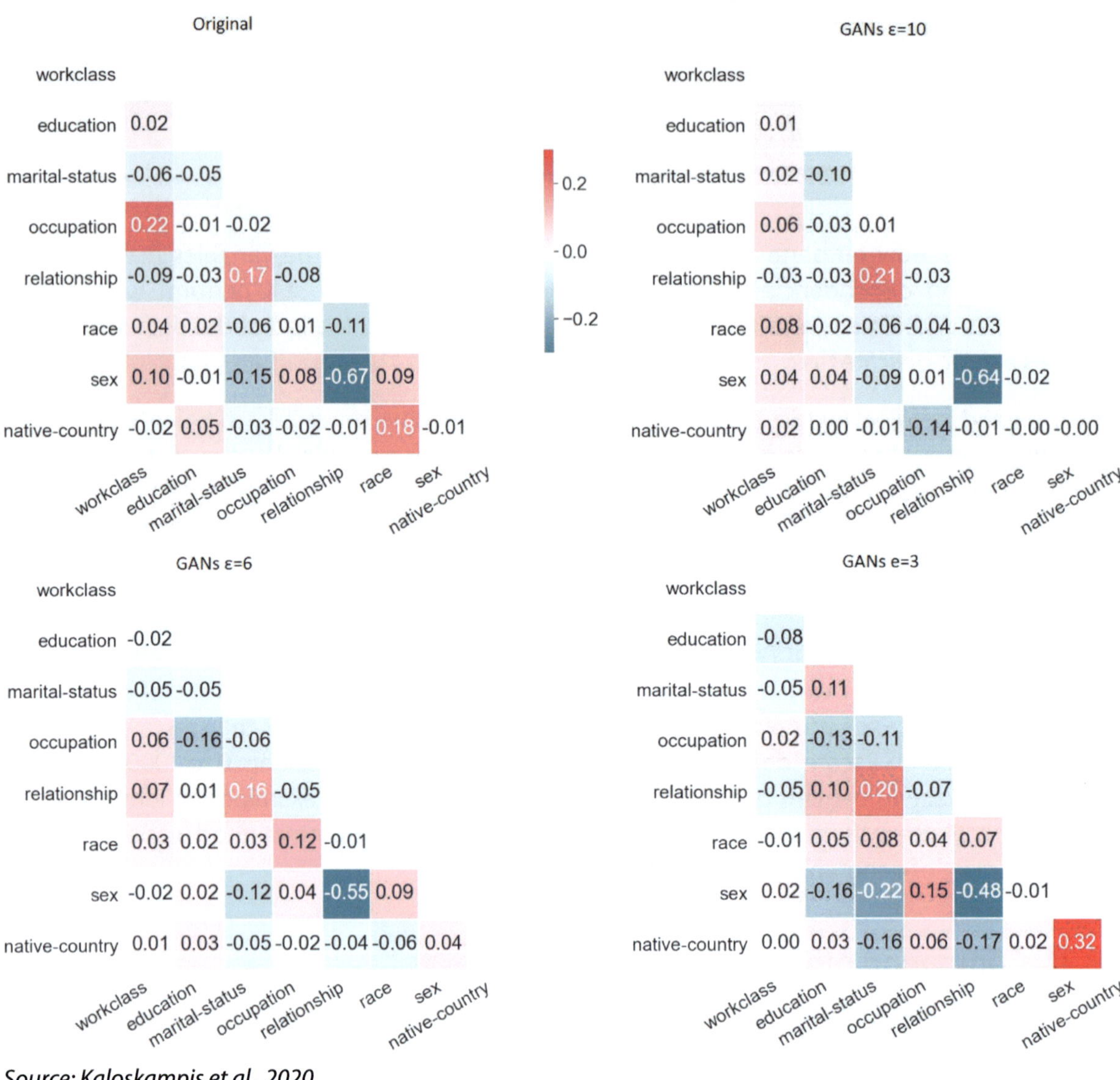

Source: Kaloskampis et al., 2020

This synthesis uses a differentially private method where the parameter ε determines the degree of privacy loss that increases as ε decreases. This method provides a means to quantify the comparison. This could be an alternative to examining two-way marginals. The marginals have the disadvantage of ignoring the ordering of variables in the utility measure. However, marginals have the advantage of being defined for categorical data as well as continuous or ordered variables.

5.4 Tips for getting started

When getting started with evaluating synthetic data based on utility, start simple. Methods such as comparing univariate distributions or task accuracy (if the task is known) between the synthetic and original data can determine a primary baseline. From there, keep in mind why you want to measure utility. Is it to compare synthesis methods or is it to improve your synthesis, often referred to as tuning? Figure 16 illustrates some of the decisions that can help to choose types of utility methods to use.

Figure 16 Utility measure decision tree

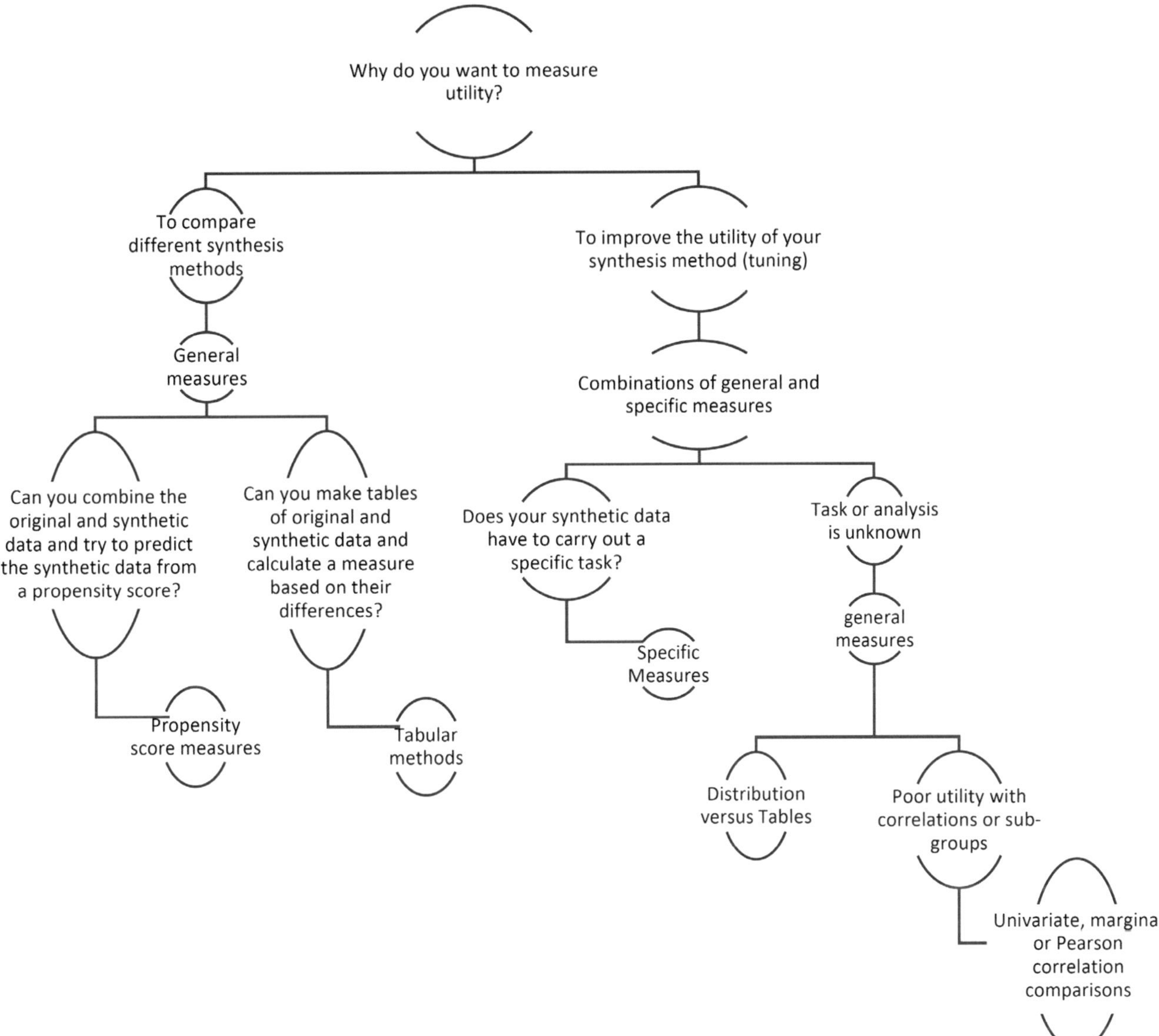

After you have determined why you want to measure utility, keep in mind your end-user and the requirements of your final synthetic data file. If the synthetic data is for a specific task or type of analysis, then specific utility measures can be the first choice. General utility measures can be useful for both comparing synthesis methods and for tuning a synthesis in respect of deficiencies identified. For the latter, marginal comparisons based on propensity scores can be useful to fine-tune your synthesis.

Keep in mind that although a large number of general utility measures have been suggested, some are equivalent to each other, and all appear to be highly correlated when compared across different data sets. For measures derived from discriminating between the synthetic and the original data, via a propensity score, the method of discrimination is more important than the utility measure chosen. We argue that a utility measure that provides only a single number is not useful in tuning the synthesis method to improve utility.

5.4.1 Some helpful equations

Table 25 presents equations for some of the more common utility measures used in practice (Raab *et al.*, 2021).

Table 25 Formulas for select utility measures

Method	Formula	Variable definitions
pMSE	$$pMSE = \sum_i (\hat{p}_i - c)^2 / N$$	$\hat{p}_i$ is the predicted probabilities, i=1,…,N, that row comes from synthetic data C is the proportion of synthesised rows in the combined data
SPECKS	$$SPECKS = \sup_{\hat{p}} |F_{t=0}(\hat{p}_i) - F_{t=1}(\hat{p}_i)|$$	n_1, n_2 and N are the number of records in the original, synthetic and combined data, respectively
PO50	$$PO50 = 100 \frac{\sum_i [t_i(\hat{p}_i > c) + (1 - t_i)(\hat{p}_i < c)]}{\sum_i (\hat{p}_i \neq c)} - 50$$	t_i indicator variable, i=1,…,N, taking the value 1 for rows form synthetic data and 0 for rows from original data y_i the counts for any original table with k categories
MabsDD	$$MabsDD = \frac{\sum_{i=1}^{k} |\frac{s_i}{n_1} - \frac{y_i}{n_2}|}{k}$$	s_i the counts for any synthetic table with k categories k total number of cells in tables to be compared

Source: Raab et al., 2021.

5.4.2 Some helpful tools

Table 26 highlights two common available and used open-source tools to calculate utility measures highlighted in this chapter.

Table 26 Common open-source packages to evaluate utility

Method	Tool
Task accuracy for deep learning models	Synthetic Data Vault (https://sdv.dev/)
pMSE	
SPECKS	*synthpop* (https://www.synthpop.org.uk/)
PO50	
MabsDD	

In addition, various utility measures were explored in the *HLG-MOS Synthetic Data Challenge 2022* (Bhagat *et al.*, 2022). The utility measures summary and results website[22] of the challenge provides a list of the utility measures and some accompanying open-source packages that were used to evaluate synthetic data utility during the challenge (Bhagat *et al.*, 2022).

---

22   See https://pages.nist.gov/HLG-MOS_Synthetic_Data_Test_Drive/index.html#utility_evaluation_methods
     for the summary and results of the utility measures evaluated in the HLG-MOS Synthetic Data Challenge 2022.

References

Abadi, M., Chu, A., Goodfellow, I., McMahan, H., Mironov, I., Talwar, K., and Zhang, L. (2016). *Deep Learning with Differential Privacy*. Proceedings of the 2016 ACM SIGSAC Conference on Computer and Communications Security (ACM CCS), pp. 308-318, 2016. arXiv:1607.00133

Abowd, J. (2016). *How Will Statistical Agencies Operate When All Data Are Private?* Washington Statistical Society Julius Shiskin Memorial Award Seminar, United States of America.

Abowd, J. (2021a). 2010 Declaration of John Abowd, State of Alabama v. United States Department of Commerce. Case No. 3:21-CV-211-RAH-ECM-KCN. (2021).

Abowd, J. (2021b). 2010 Supplemental Declaration of John M. Abowd, State of Alabama v. United States Department of Commerce. Case No. 3:21-CV-211-RAH-ECM-KCN. (2021).

Basu, D. (1971). *An essay on the logical foundations of survey sampling, part 1. In Foundations of Statistical Inference*. Edited by V.P. Godambe and D.A. Sprott. 203-242. Toronto: Holt, Rinehart and Winston.

Beaulieu-Jones, B., Wu, S., Williams, C., Lee, R., Bhavnani, S., Byrd, J., and Greene, C. (2019). *Privacy-Preserving Generative Deep Neural Networks Support Clinical Data Sharing*. Circulation: Cardiovascular Quality and Outcomes. Volume 12, Issue 7, July 2019 https://doi.org/10.1161/CIRCOUTCOMES.118.005122.

Bhagat, K., Task, C., Howarth, G., Wall, M. and Burnett-Isaacs, K. (2022, August 22). *HLG-MOS Synthetic Data Test-Drive 2022*. NIST. Retrieved August 22, 2022, from https://pages.nist.gov/HLG-MOS_Synthetic_Data_Test_Drive/

Bhattacharyya, A. (1943). *On a measure of divergence between two statistical populations defined by their probability distributions*. Calcutta Mathematical Society, 35, 99–109.

Bowen, C.M., and Lui, F. and Su, B. (2021). *Differentially private data release via statistical election to partition sequentially*. METRON, 79(1), 1–31. URL https://doi.org/10.1007/s40300-021-00201-0.

Bowen, C. and Snoke, J. (2021). *Comparative Study of Differentially Private Synthetic Data Algorithms from the NIST PSCR Differential Privacy Synthetic Data Challenge*. Journal of Privacy and Confidentiality 11 (1).

Cano, I. and Torra, V. (2009). *Generation of Synthetic Data by means of fuzzy c-Regression*. 1145-1150. 10.1109/FUZZY.2009.5277074.

Chawla, N. V., Bowyer, K. W., Hall, L. O. and Kegelmeyer, W. P. (2002). *SMOTE: Synthetic Minority Over-sampling Technique*. Journal of Artificial Intelligence Research, 16, 321–357.

Data, Responsibly (November 15, 2021). *DataSynthesizer*. Retrieved March 23, 2022 https://pypi.org/project/DataSynthesizer/.

Desai, T., Ritchie, F. and Welpton, R. (2016). *Fives Safes: Designing Data Access for Research*. University of the West England Research Repository.

Domingo-Ferrer, J. and Gonzalez-Nicolas, U. (2010). *Hybrid microdata using microaggregation.* Information Sciences, 180(15), 2834-2844.

Dosselmann, R., Sadeqi, M. and Hamilton, H. J. (2019). *A Tutorial on Computing t-Closeness.* arXiv preprint arXiv:1911.11212.

Drechsler, J. (2011). *Synthetic Data Sets for Statistical Disclosure Control.* Springer, New York.

Drechsler, J. and Reiter, J.P. (2009). *Disclosure Risk and Data Utility for Partially Synthetic Data: An Empirical Study Using the German IAB Establishment Survey.* Journal of Official Statistics, Vol. 25, No. 4, pp 589-603.

Drechsler, J. and Reiter, J.P. (2011). *An empirical evaluation of easily implemented, nonparametric methods for generating synthetic datasets.* Computational Statistics and Data Analysis, 55, 3232-3243.

Duncan, G., Elliot, M. and Salazar-Gonzalez, J. (2011). *Statistical Confidentiality: Principles and practice.* Springer, New-York.

Dwork, C., McSherry, F., Nissim, K. and Smith, A. (2006). *Calibrating noise to sensitivity in private data analysis.* Proceedings of the 3rd Theory of Cryptography Conference, 265-284.

Efron, B. (1979). *Bootstrap Methods: Another Look at the Jackknife.* Ann. Statist. 7 (1) 1 – 26, January.

El Emam, K. (2013). *Guide to the De-Identification of Personal Health Information.* CRC Press.

Elliot, M., Mackey, E. and O'Hara., K. (2016). *The Anonymization Decision Making Framework.* UK Anonymization Network, Manchester.

Eurostat (2007*), European Statistical System Code of Practice Peer Reviews: The National Statistical Institute's guide.* Luxembourg. https://unstats.un.org/unsd/dnss/docs-nqaf/Eurostat-Peer%20 review%20NSI%20guide.pdf

Fekri, M.N., Ghosh, A.M. and Grolinger, K. (2020). *Generating energy data for machine learning with recurrent generative adversarial networks.* Energies 13, 1 (2020), 130.

Fleishman, A. (1978). *A Method for Simulating Non-Normal Distributions.*

Fuglede, B. and Topsoe, F. (2004). *Jensen-Shannon divergence and Hilbert space embedding* (PDF). Proceedings of the International Symposium on Information Theory, 2004. IEEE. p. 30. doi:10.1109/ ISIT.2004.1365067. ISBN 978-0-7803-8280-0. S2CID 7891037

Goodfellow, I., Pouget-Abadie, J., Mirza, M., Xu, B., Warde-Farley, D., Ozair, S., Courville, A. and Bengio, Y. (2014). *Generative Adversarial Networks.* Advances in Neural Information Processing Systems. 3. 10.1145/3422622.

Hawes, M.B. (2020). *Implementing Differential Privacy: Seven Lessons From the 2020 United States Census.*

Heyburn, R., Bond, R.R., Black, M., Mulvenna, M., Wallace, J., Rankin, D. and Cleland, B. (2018). *Machine learning using synthetic and real data: similarity of evaluation metrics for different healthcare datasets and for different algorithms*. In Data Science and Knowledge Engineering for Sensing Decision Support: Proceedings of the 13th International FLINS Conference (FLINS 2018). World Scientific, 1281–1291.

Jordon, J., Yoon, J. and Van Der Schaar, M. (2018). *Measuring the quality of synthetic data for use in competitions*. arXiv preprint arXiv:1806.11345 (2018).

Jorgensen, T. D., Pornprasertmanit, S., Schoemann, A. M. and Rosseel, Y. (2019). *semTools: Useful Tools for Structural Equation Modeling*.

Kaloskampis, I., Pugh, D., Joshi, C. and Nolan, L., *Synthetic data for public good*, ONS Data Science Campus blog, 2019. https://datasciencecampus.ons.gov.uk/projects/synthetic-data-for-public-good/

Kaloskampis, I., Joshi, C., Cheung, C., Pugh, D. and Nolan, L. (2020). *Synthetic data in the civil service*. Significance, 17: 18-23. https://doi.org/10.1111/1740-9713.01466

Karr, A. F., Kohnen, C. N., Oganian, A., Reiter, J. P. and Sanil, A. P. (2006). *A framework for evaluating the utility of data altered to protect confidentiality*. The American statistician, 60(3):224{232.

Kim, H. J., Drechsler, J. and Thompson, K. J. (2020). *Synthetic microdata for establishment surveys under informative sampling*. Journal of the Royal Statistical Society Series A, Royal Statistical Society, vol. 184(1), pages 255-281, January.

Kohavi, R. and Becker, B. (1996). *US Adult Income dataset*. UCI Machine Learning Repository. bit.ly/3dcnUmZ.

Langsrud, Ø. (2019). *Information Preserving Regression-based Tools for Statistical Disclosure Control*. Statistics and Computing, 29, 965–976.

Lavallée, P. and Beaumont, J.-F. (2015). *Why We Should Put Some Weight on Weights. Survey Insights: Methods from the Field, Weighting: Practical Issues and 'How to' Approach*. Invited article, Retrieved from https://surveyinsights.org/?p=6255

L'Écuyer, P. and Puchhammer, F. (2021). *Density Estimation by Monte Carlo and Quasi-Monte Carlo*. Monte Carlo and Quasi-Monte Carlo Methods.

Leduc, J. and Grislain, N. (2021). *Composable Generative Models*. arXiv: 2102.09249v1 https://arxiv.org/pdf/2102.09249.pdf

LeFevre, K., DeWitt, D. and Ramakrishnan, R. (2005). *Incognito: Efficient full-domain K-anonymity*. Paper presented at the 49-60. https://doi.org/10.1145/1066157.1066164

Li, N., Li, T. and Venkatasubramanian, S. (2007). *T-Closeness: Privacy Beyond k-Anonymity and ℓ-Diversity. t-Closeness: Privacy beyond k-anonymity and ℓ-Diversity* (PDF). pp. 106–115. doi:10.1109/ICDE.2007.367856

Lohr, S. (1999). Sampling: Design and Analysis. Duxbury Press.

Machanavajjhala, A., Kifer, D., Gehrke, J. and Venkitasubramaniam, M. (2007). *-Diversity: Privacy beyond k-anonymity*. ACM Trans. Knowl. Discov. Data 1, 1, Article 3 (March 2007), 52 pages. DOI = 10.1145/1217299.1217302 http://doi.acm.org/10.1145/1217299.1217302

McKenna, R., Sheldon, D. and Miklau, G. (2019). *Graphical-model based estimation and inference for differential privacy*. In Proceedings of the 36th International Conference on Machine Learning.

Mendes, R. and Vilela, J. P. (2017). *Privacy-preserving data mining: methods, metrics, and applications*. IEEE Access, 5, 10562-10582.

Muralidhar, K. and Sarathy, R. (2008). *Generating Sufficiency-based Non-synthetic Perturbed Data*. Transactions on Data Privacy, 1(1), 17-33.

New Zealand Income Survey Super SURF. (n.d.). *New Zealand Income Survey Super SURF*; web. archive.org. Retrieved August 21, 2022, from https://web.archive.org/web/20201021153827/ http:/archive.stats.govt.nz/tools_and_services/university-students/NZIS-Super-SURF.aspx #gsc.tab=0

NIST. (2021, January 7). *2018 Differential Privacy Synthetic Data Challenge*. NIST. Retrieved October 20, 2021, from https://www.nist.gov/ctl/pscr/open-innovation-prize challenges/past-prize-challenges/2018-differential-privacy-synthetic

Nowok, B., Raab, G. M. and Dibben, C. (2015). *synthpop: Bespoke creation of synthetic data in R*. Package vignette http://cran.r-project.org/web/packages/synthpop/vignettes/synthpop.pdf. Accessed: 2015-02-26.

Nowok, B., Dibben, C. and Raab, G. (2017). *Recognising real people in synthetic microdata: risk mitigation and impact on utility*. Paper presented to UNECE Work Session on Statistical Data Confidentiality, Skopje North Macedonia, Available from https://unece.org/statistics/events/ SDC2017

OECD (2003). *The OECD Glossary of Statistical Terms*. Retrieved August 21, 2022 from https://stats.oecd.org/glossary/

Olkin, I. (1987). *A Conversation with Morris Hansen*. Statistical Science. Vol.2, No.2, pp. 162-179.

Raab, G.M. and Nowok, B. (2017). *Inference from fitted models in synthpop*. R Vignette, https://cran.r-project.org/web/packages/synthpop/vignettes/inference.pdf

Raab, G. M., Nowok, B. and Dibben, C. (2021). *Assessing, visualizing and improving the utility of synthetic data*. Paper submitted to UNECE Work Session on Statistical Data Confidentiality 2021' Available from https://arxiv.org/pdf/2109.12717.pdf

Rancourt, E. (2019). *The scientific approach as a transparency enabler throughout the data life-cycle*. Statistical Journal of the IAOS 35, 549-558.

Read, T. and Cressie, R.C. (1988). *Goodness-of-Fit Statistics for Discrete Multivariate Data*. NAC, Springer, Berlin.

Reiter, J.P. (2005). *Using CART to generate partially synthetic public use microdata.* Journal of Official Statistics; 21(3): pp 441–462.

Reiter, J. and Mitra, R. (2009). *Estimating Risks of Identification Disclosure in Partially Synthetic Data.* The Journal of Privacy and Confidentiality 1, Number 1, pp. 99–110 https://journalprivacyconfidentiality.org/index.php/jpc/article/view/567/550

Ridgeway, D., Theofanos, M., Manley, T. and Task, C. (2021). *Challenge Design and Lessons Learned from the 2018 Differential Privacy Challenges.* Technical Note (NIST TN), National Institute of Standards and Technology, Gaithersburg, MD, [online], https://doi.org/10.6028/NIST.TN.2151, https://tsapps.nist.gov/publication/get_pdf.cfm?pub_id=931343 (Accessed October 20, 2021)

Ruggles, S., Flood, S., Goeken, R., Grover, J., Meyer, E., Pacas, J. and Sobek, M. PUMS USA: *Version 8.0 Extract of 1940 Census for U.S. Census Bureau Disclosure Avoidance Research [dataset].* Minneapolis, 2018.

Sallier, K. (2020). *Toward More User-centric Data Access Solutions: Producing Synthetic Data of High Analytical Value by Data Synthesis.* Statistical Journal of the IAOS, vol. 36, no. 4, pp. 1059-1066.

Sallier, K. and Girard, C. (2018). *Toward a Successful Implementation of Synthesis in a National Statistical Agency: A Model for Cooperation.* Privacy in Statistical Databases, conference held in Valencia, Spain.

Slokom, M. and Larson, M. (2021). *Doing Data Right: How Lessons Learned Working with Conventional Data should Inform the Future of Synthetic Data for Recommender Systems.* Simulation and synthetic data for recommender systems (SimuRec) Workshop, in conjunction with the 15th ACM Conference on Recommender Systems.

Snoke, J., Raab, G. M., Nowok, B., Dibben, C. and Slavkovic, A. (2018). *General and specific utility measures for synthetic data.* Journal of the Royal Statistical Society. Series A: Statistics in Society., 181:663{668.

Soria-Comas, J., Domingo-Ferrer, J., Sanchez, D. and Martinez, S. (2015). *t-closeness through microaggregation: Strict privacy with enhanced utility preservation.* IEEE Transactions on Knowledge and Data Engineering, 27(11), 3098-3110.

Statistics Canada (2003). *Survey methods and practices.* Statistics Canada Catalogue no. 12-587-XPE, Ottawa, Ontario. 396 p.

Sweeney, L. (2002). *k-anonymity: A model for protecting privacy.* International Journal of Uncertainty, Fuzziness, and Knowledge-Based Systems, 10(5), 557-570. https://doi.org/10.1142/S0218488502001648

Ting, D., Fienberg, S.E. and Trottini, M. (2008). *Random orthogonal matrix masking methodology for microdata release.* International Journal of Information and Computer Security, 2(1), 86-105.

Van Buuren, S., Brand, J. P. L., Groothuis-Oudshoorn, C. G. M. and Rubin, D. B. (2006). *Fully Conditional Specification in Multivariate Imputation.* Journal of Statistical Computation and Simulation 76 (12): 1049–64.

Vale, C. D. and Maurelli, V. A. (1983). *Simulating Multivariate Nonnormal Distributions*. Psychometrika 48 (3): 465–71.

Voas, D. and Williamson, P. (2001). *Evaluating goodness-of-fit measures for synthetic microdata*. Geographical and Environmental Modelling, 5(2).

Wasserman, L. and Zhou, S. (2010). *A Statistical Framework for Differential Privacy*. Journal of the American Statistical Association, 105:489, 375-389, DOI: 10.1198/jasa.2009.tm08651

Woo, M.J., Reiter, J.P., Oganian, A. and Karr, A.F. (2009). *Global Measures of Data Utility for Microdata Masked for Disclosure Limitation*. Journal of Privacy and Confidentiality, 1, 111–124.